The Complete U.S. Army Survival Guide to

FIRECRAFT

Skills, Tactics, And Techniques

The Complete U.S. Army Survival Guide to
FIRECRAFT
Skills, Tactics, And Techniques

Edited by

Jay McCullough

Skyhorse Publishing

Skyhorse Publishing books may be purchased in bulk at special discounts for sales promotion, corporate gifts, fund-raising, or educational purposes. Special editions can also be created to specifications. For details, contact the Special Sales Department, Skyhorse Publishing, 307 West 36th Street, 11th Floor, New York, NY 10018 or info@skyhorsepublishing.com.

Skyhorse® and Skyhorse Publishing® are registered trademarks of Skyhorse Publishing, Inc.®, a Delaware corporation.

Visit our website at www.skyhorsepublishing.com.

10 9 8 7 6 5 4 3 2

Library of Congress Cataloging-in-Publication Data is available on file.

Cover design by Tom Lau

Print ISBN: 978-1-5107-0744-3
Ebook ISBN: 978-1-5107-0749-8

Printed in China

Table of Contents

Introduction

Just as there is no struggle as primitive and desperate as hand-to-hand combat, there is no substitute for practical combat training. In this book of materials culled from various U.S. Army publications, you will find not only an excellent introduction to this discipline, but also a valuable resource for those whose skills have become rusty. This book is especially notable because it is drawn from decades of real combat experience; it is a distillation of techniques that work, tips to avoid potential problems, and eschews poor practices that would put a soldier at a fatal disadvantage.

So too, would it be a serious disadvantage not to know how to seek and maintain cover, create camouflage, analyze the position and disposition of an enemy through tracking skills, conduct basic navigation, and structure a scouting party in day or night. Here you will find the basics as described by the U.S. Army as the necessary knowledge for foot soldiers.

Here also you will find simple, timeless field expedients for starting fires, building elementary weapons, and storing and cooking food. Practicality is perhaps this manual's strongest suit: It was prepared by the Army for Everyman—or woman—to be near-universal, easy-to-comprehend, no-nonsense, and inclusive of methods that have passed the test of time and that are in some cases more ancient than our own species.

Similarly, to begin where we started: There is no substitute for practical training. If we would be prepared for any eventuality, as the Army has seen to do with its training manuals, it would be wise to commit these techniques to memory, and to practice them as necessary.

—Jay McCullough
April 2016
North Haven, Connecticut

CHAPTER 1

FIRECRAFT

In many survival situations, the ability to start a fire can make the difference between living and dying. Fire can fulfill many needs. It can provide warmth and comfort. It not only cooks and preserves food, it also provides warmth in the form of heated food that saves calories our body normally uses to produce body heat. You can use fire to purify water, sterilize bandages, signal for rescue, and provide protection from animals. It can be a psychological boost by providing peace of mind and companionship. You can also use fire to produce tools and weapons.

Fire can cause problems, as well. The enemy can detect the smoke and light it produces. It can cause forest fires or destroy essential equipment. Fire can also cause burns and carbon monoxide poisoning when used in shelters.

Remember weigh your need for fire against your need to avoid enemy detection.

BASIC FIRE PRINCIPLES

To build a fire, it helps to understand the basic principles of a fire. Fuel (in a nongaseous state) does not burn directly. When you apply heat to a fuel, it produces a gas. This gas, combined with oxygen in the air, burns.

Understanding the concept of the fire triangle is very important incorrectly constructing and maintaining a fire. The three sides of the triangle represent air, heat, and fuel. If you remove any of these, the fire will go out. The correct ratio of these components is very important for a fire to burn at its greatest capability. The only way to learn this ratio is to practice.

SITE SELECTION AND PREPARATION

You will have to decide what site and arrangement to use. Before building a fire consider—

- The area (terrain and climate) in which you are operating.
- The materials and tools available. Time: how much time you have?
- Need: why you need a fire?
- Security: how close is the enemy?

Look for a dry spot that—

- Is protected from the wind.
- Is suitably placed in relation to your shelter (if any).
- Will concentrate the heat in the direction you desire.
- Has a supply of wood or other fuel available. (See Table 1-1 for types of material you can use.)

If you are in a wooded or brush-covered area, clear the brush and scrape the surface soil from the spot you have selected. Clear a circle at least 1 meter in diameter so there is little chance of the fire spreading.

If time allows, construct a fire wall using logs or rocks. This wall will help to reflector direct the heat where you want it (Figure 1-1). It will also reduce flying sparks and cut down on the amount of wind blowing into the fire. However, you will need enough wind to keep the fire burning.

> ⚠ **CAUTION**
> Do not use wet or porous rocks as they may explode when heated.

In some situations, you may find that an underground fireplace will best meet your needs. It conceals the fire and serves well for cooking food. To make an underground fireplace or Dakota fire hole (Figure 1-2)—

- Dig a hole in the ground.
- On the upwind side of this hole, poke or dig a large connecting hole for ventilation.
- Build your fire in the hole as illustrated.

Table 1-1: Materials for Building Fires

Tinder	Kindling	Fuel
• Birch bark • Shredded inner bark from cedar, chestnut, red elm trees • Fine wood shavings • Dead grass, ferns, moss, fungi • Straw • Sawdust • Very fine pitchwood scrapings • Dead evergreen needles • Punk (the completely rotted portions of dead logs or trees) • Evergreen tree knots • Bird down (fine feathers) • Down seed heads (milkweed, dry cattails, bulrush, or thistle) • Fine, dried vegetable fibers • Spongy threads of dead puffball • Dead palm leaves • Skinlike membrane lining bamboo • Lint from pocket and seams • Charred cloth • Waxed paper • Outer bamboo shavings • Gunpowder • Cotton • Lint	• Small twigs • Small strips of wood • Split wood • Heavy cardboard • Pieces of wood removed from the inside of larger pieces • Wood that has been doused with highly flammable materials, such as gasoline, oil, or wax	• Dry, standing wood and dry, dead branches • Dry inside (heart) of fallen tree trunks and large branches • Green wood that is finely split • Dry grasses twisted into bunches • Peat dry enough to burn (this may be found at the top of undercut banks) • Dried animal dung • Animal fats • Coal, oil shale, or oil lying on the surface

Figure 1-1: Types of fire walls

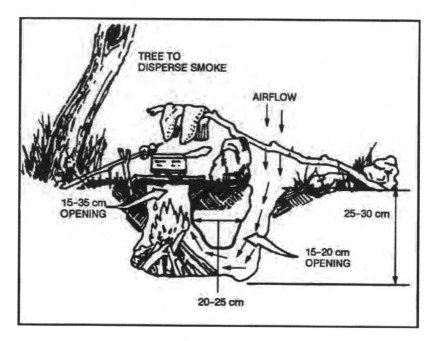

Figure 1-2: Dakota fire hole

If you are in a snow-covered area, use green logs to make a dry base for your fire (Figure 1-3). Trees with wrist-sized trunks are easily broken in extreme cold. Cut or break several green logs and lay them side by side on top of the snow. Add one or two more layers. Lay the top layer of logs opposite those below it.

Figure 1-3: Base for fire in snow-covered area

FIRE MATERIAL SELECTION

You need three types of materials (Table 1-1) to build a fire—tinder, kindling, and fuel.

Tinder is dry material that ignites with little heat —a spark starts a fire. The tinder must be absolutely dry to be sure just a spark will ignite it. If you only have a device that generates sparks, charred cloth will be almost essential. It holds a spark for long periods, allowing you to put tinder on the hot area to generate a small flame. You can make charred cloth by heating cotton cloth until it turns black, but does not burn. Once it is black, you must keep it in an airtight container to keep it dry. Prepare this cloth well in advance of any survival situation. Add it to your individual survival kit.

Kindling is readily combustible material that you add to the burning tinder. Again, this material should be absolutely dry to ensure rapid burning. Kindling increases the fire's temperature so that it will ignite less combustible material.

Fuel is less combustible material that burns slowly and steadily once ignited.

HOW TO BUILD A FIRE

There are several methods for laying a fire, each of which has advantages. The situation you find yourself in will determine which fire to use.

Tepee

To make this fire (Figure 1-4), arrange the tinder and a few sticks of kindling in the shape of a tepee or cone. Light the center. As the tepee burns, the outside logs will fall inward, feeding the fire. This type of fire burns well even with wet wood.

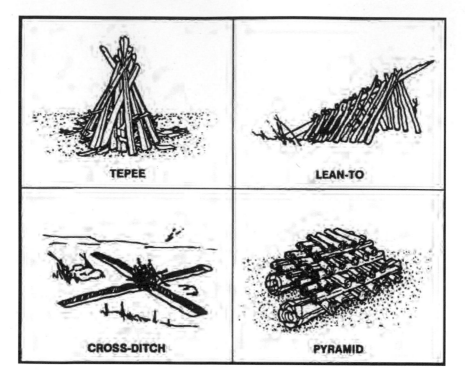

Figure 1-4: Methods for laying fires

Lean-To

To lay this fire (Figure 1-4), push a green stick into the ground at a 30-degree angle. Point the end of the stick in the direction of the wind. Place some tinder deep under this lean-to stick. Lean pieces of kindling against the lean-to stick. Light the tinder. As the kindling catches fire from the tinder, add more kindling.

Cross-Ditch

To use this method (Figure 1-4), scratch a cross about 30 centimeters in size in the ground. Dig the cross 7.5 centimeters deep. Put a large wad of tinder in the middle of the cross. Build a kindling pyramid above the tinder. The shallow ditch allows air to sweep under the tinder to provide a draft.

Pyramid

To lay this fire (Figure 1-4), place two small logs or branches parallel on the ground. Place a solid layer of small logs across the parallel logs. Add three or four more layers of logs or branches, each layer smaller than and at a right angle to the layer below it. Make a starter fire on top of the pyramid. As the starter fire burns, it will ignite the logs below it. This gives you a fire that burns downward, requiring no attention during the night.

There are several other ways to lay a fire that are quite effective. Your situation and the material available in the area may make another method more suitable.

HOW TO LIGHT A FIRE

Always light your fire from the upwind side. Make sure to lay your tinder, kindling, and fuel so that your fire will burn as long as you need it. Igniters provide the initial heat required to start the tinder burning. They fall into two categories: modem methods and primitive methods.

Modern Methods

Modem igniters use modem devices—items we normally think of to start a fire.

Matches

Make sure these matches are waterproof. Also, store them in a waterproof container along with a dependable striker pad.

Convex Lens

Use this method (Figure 1-5) only on bright, sunny days. The lens can come from binoculars, camera, telescopic sights, or magnifying glasses. Angle the lens to concentrate the sun's rays on the tinder. Hold the lens over the same spot until the tinder begins to smolder. Gently blow or fan the tinder into flame, and apply it to the fire lay.

Metal Match

Place a flat, dry leaf under your tinder with a portion exposed. Place the tip of the metal match on the dry leaf, holding the metal match in one hand and a knife in the other. Scrape your knife against the metal match to produce sparks. The sparks will hit the tinder. When the tinder starts to smolder, proceed as above.

Battery

Use a battery to generate a spark. Use of this method depends on the type of battery available. Attach a wire to each terminal. Touch the

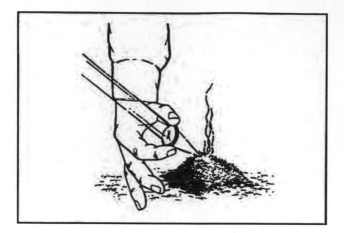

Figure 1-5: Lens method

ends of the bare wires together next to the tinder so the sparks will ignite it.

Gunpowder

Often, you will have ammunition with your equipment. If so, carefully extract the bullet from the shell casing, and use the gunpowder as tinder. A spark will ignite the powder. Be extremely careful when extracting the bullet from the case.

Primitive Methods

Primitive igniters are those attributed to our early ancestors.

Flint and Steel

The direct spark method is the easiest of the primitive methods to use. The flint and steel method is the most reliable of the direct spark methods. Strike a flint or other hard, sharp-edged rock edge with a piece of carbon steel (stainless steel will not produce a good spark). This method requires a loose-jointed wrist and practice. When a spark has caught in the tinder, blow on it. The spark will spread and burst into flames.

Fire-Plow

The fire-plow (Figure 1-6) is a friction method of ignition. You rub a hardwood shaft against a softer wood base. To use this method, cut a straight groove in the base and plow the blunt tip of the shaft up and down the groove. The plowing action of the shaft pushes out small particles of wood fibers. Then, as you apply more pressure on each stroke, the friction ignites the wood particles.

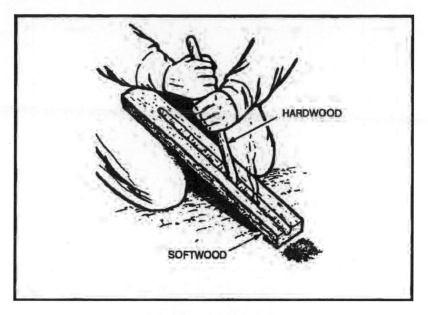

Figure 1-6: Fire-plow

Bow and Drill

The technique of starting a fire with a bow and drill (Figure 1-7) is simple, but you must exert much effort and be persistent to produce a fire. You need the following items to use this method:

- Socket. The socket is an easily grasped stone or piece of hard-wood or bone with a slight depression in one side. Use it to hold the drill in place and to apply downward pressure.
- Drill. The drill should be a straight, seasoned hardwood stick about 2 centimeters in diameter and 25 centimeters long. The top end is round and the low end blunt (to produce more friction).
- Fire board. Its size is up to you. A seasoned softwood board about 2.5 centimeters thick and 10 centimeters wide is preferable. Cut a depression about 2 centimeters from the edge on one side of the board. On the underside, make a V-shaped cut from the edge of the board to the depression.
- Bow. The bow is a resilient, green stick about 2.5 centimeters in diameter and a string. The type of wood is not important. The bowstring can be any type of cordage. You tie the bow-string from one end of the bow to the other, without any slack.

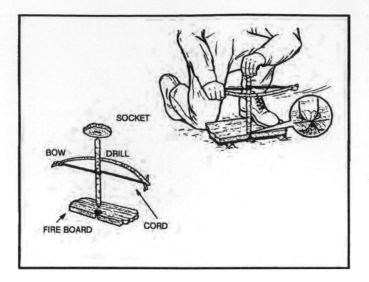

Figure 1-7: Bow and drill

To use the bow and drill, first prepare the fire lay. Then place a bundle of tinder under the V-shaped cut in the fire board. Place one foot on the fire board. Loop the bowstring over the drill and place the drill in the precut depression on the fire board. Place the socket, held in one hand, on the top of the drill to hold it in position. Press down on the drill and saw the bow back and forth to twirl the drill (Figure 1-7). Once you have established a smooth motion, apply more downward pressure and work the bow faster. This action will grind hot black powder into the tinder, causing a spark to catch. Blow on the tinder until it ignites.

Note: Primitive fire-building methods are exhaustive and require practice to ensure success.

Helpful Hints
- Use nonaromatic seasoned hardwood for fuel, if possible.
- Collect kindling and tinder along the trail.
- Add insect repellent to the tinder.
- Keep the firewood dry.
- Dry damp firewood near the fire.
- Bank the fire to keep the coals alive overnight,
- Carry lighted punk, when possible.
- Be sure the fire is out before leaving camp.
- Do not select wood lying on the ground. It may appear to be dry but generally doesn't provide enough friction.

CHAPTER 2

FIELD-EXPEDIENT WEAPONS, TOOLS, AND EQUIPMENT

As a soldier you know the importance of proper care and use of your weapons, tools, and equipment. This is especially true of your knife. You must always keep it sharp and ready to use. A knife is your most valuable tool in a survival situation. Imagine being in a survival situation without any weapons, tools, or equipment except your knife. It could happen! You might even be without a knife. You would probably feel helpless, but with the proper knowledge and skills, you can easily improvise needed items.

In survival situations, you may have to fashion any number and type of field-expedient tools and equipment to survive. Examples of tools and equipment that could make your life much easier are ropes, rucksacks, clothes, nets, and so on.

Weapons serve a dual purpose. You use them to obtain and prepare food and to provide self-defense. A weapon can also give you a feeling of security and provide you with the ability to hunt on the move.

CLUBS

You hold clubs, you do not throw them. As a field-expedient weapon, the club does not protect you from enemy soldiers. It can, however, extend your area of defense beyond your fingertips. It also serves to increase the force of a blow without injuring yourself. There are three basic types of clubs. They are the simple, weighted, and sling club.

Simple Club
A simple club is a staff or branch. It must be short enough for you to swing easily, but long enough and strong enough for you to damage

whatever you hit. Its diameter should fit comfortably in your palm, but it should not be so thin as to allow the club to break easily upon impact. A straight-grained hardwood is best if you can find it.

Weighted Club

A weighted club is any simple club with a weight on one end. The weight may be a natural weight, such as a knot on the wood, or something added, such as a stone lashed to the club.

To make a weighted club, first find a stone that has a shape that will allow you to lash it securely to the club. A stone with a slight hourglass shape works well. If you cannot find a suitably shaped stone, you must fashion a groove or channel into the stone by a technique known as pecking. By repeatedly rapping the club stone with a smaller hard stone, you can get the desired shape.

Next, find a piece of wood that is the right length for you. A straight-grained hardwood is best. The length of the wood should feel comfortable in relation to the weight of the stone. Finally, lash the stone to the handle.

There are three techniques for lashing the stone to the handle: split handle, forked branch, and wrapped handle. The technique you use will depend on the type of handle you choose. See Figure 2-1.

Sling Club

A sling club is another type of weighted club. A weight hangs 8 to 10 centimeters from the handle by a strong, flexible lashing (Figure 2-2). This type of club both extends the user's reach and multiplies the force of the blow.

EDGED WEAPONS

Knives, spear blades, and arrow points fall under the category of edged weapons. The following paragraphs will discuss the making of such weapons.

Knives

A knife has three basic functions. It can puncture, slash or chop, and cut. A knife is also an invaluable tool used to construct other survival items. You may find yourself without a knife or you may need another type knife or a spear. To improvise you can use stone, bone, wood, or metal to make a knife or spear blade.

Stone

To make a stone knife, you will need a sharp-edged piece of stone, a chipping tool, and a flaking tool. A chipping tool is a light, blunt-edged

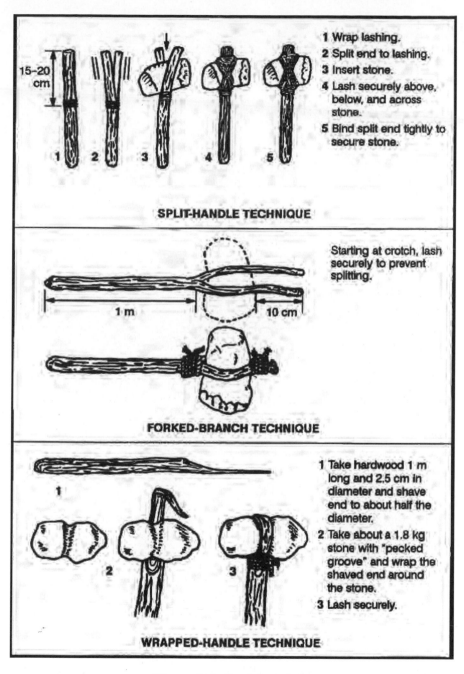

SPLIT-HANDLE TECHNIQUE

1 Wrap lashing.
2 Split end to lashing.
3 Insert stone.
4 Lash securely above, below, and across stone.
5 Bind split end tightly to secure stone.

FORKED-BRANCH TECHNIQUE

15-20 cm

1 m   10 cm

Starting at crotch, lash securely to prevent splitting.

WRAPPED-HANDLE TECHNIQUE

1 Take hardwood 1 m long and 2.5 cm in diameter and shave end to about half the diameter.
2 Take about a 1.8 kg stone with "pecked groove" and wrap the shaved end around the stone.
3 Lash securely.

Figure 2-1: Lashing clubs

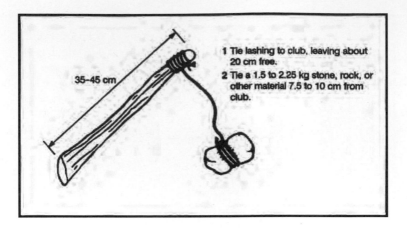

1 Tie lashing to club, leaving about 20 cm free.

2 Tie a 1.5 to 2.25 kg stone, rock, or other material 7.5 to 10 cm from club.

35–45 cm

Figure 2-2: Sling club

tool used to break off small pieces of stone. A flaking tool is a pointed tool used to break off thin, flattened pieces of stone. You can make a chipping tool from wood, bone, or metal, and a flaking tool from bone, antler tines, or soft iron (Figure 2-3).

Start making the knife by roughing out the desired shape on your sharp piece of stone, using the chipping tool. Try to make the knife fairly thin. Then, using the flaking tool, press it against the edges. This action will cause flakes to come off the opposite side of the edge, leaving a razor sharp edge. Use the flaking tool along the entire length of the edge you need to sharpen. Eventually, you will have a very sharp cutting edge that you can use as a knife.

Lash the blade to some type of hilt (Figure 2-3).

Note: Stone will make an excellent puncturing tool and a good chopping tool but will not hold a fine edge. Some stones such as chert or flint can have very fine edges.

Bone

You can also use bone as an effective field-expedient edged weapon. First, you will need to select a suitable bone. The larger bones, such as the leg bone of a deer or another medium-sized animal, are best. Lay the bone upon another hard object. Shatter the bone by hitting it with a heavy object, such as a rock. From the pieces, select a suitable pointed splinter. You can further shape and sharpen this splinter by rubbing it on a rough-surfaced rock. If the piece is too small to handle, you can still use it by adding a handle to it. Select a suitable piece of hardwood for a handle and lash the bone splinter securely to it.

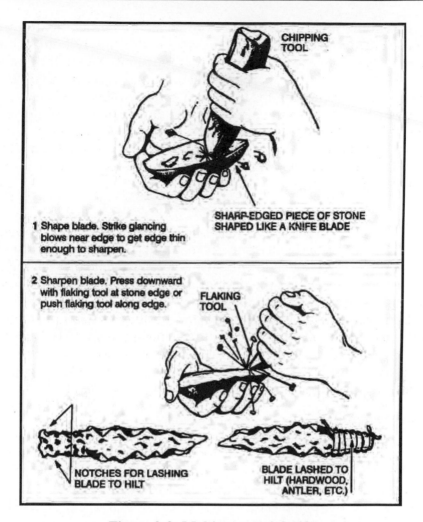

Figure 2-3: Making a stone knife

Note: Use the bone knife only to puncture. It will not hold an edge and it may flake or break if used differently.

Wood

You can make field-expedient edged weapons from wood. Use these only to puncture. Bamboo is the only wood that will hold a suitable edge. To make a knife using wood, first select a straight-grained piece of hardwood that is about 30 centimeters long and 2.5 centimeters in diameter. Fashion the blade about 15 centimeters long. Shave it down to a point. Use only the straight-grained portions of the wood. Do not use the core or pith, as it would make a weak point.

Harden the point by a process known as fire hardening. If a fire is possible, dry the blade portion over the fire slowly until lightly charred. The drier the wood, the harder the point. After lightly charring the blade portion, sharpen it on a coarse stone. If using bamboo and after fashioning the blade, remove any other wood to make the blade thinner from the inside portion of the bamboo. Removal is done this way because bamboo's hardest part is its outer layer. Keep as much of this layer as possible to ensure the hardest blade possible. When charring bamboo over a fire, char only the inside wood; do not char the outside.

Metal

Metal is the best material to make field-expedient edged weapons. Metal, when properly designed, can fulfill a knife's three uses—puncture, slice or chop, and cut. First, select a suitable piece of metal, one that most resembles the desired end product. Depending on the size and original shape, you can obtain a point and cutting edge by rubbing the metal on a rough-surfaced stone. If the metal is soft enough, you can hammer out one edge while the metal is cold. Use a suitable flat, hard surface as an anvil and a smaller, harder object of stone or metal as a hammer to hammer out the edge. Make a knife handle from wood, bone, or other material that will protect your hand.

Other Materials

You can use other materials to produce edged weapons. Glass is a good alternative to an edged weapon or tool, if no other material is available. Obtain a suitable piece in the same manner as described for bone. Glass has a natural edge but is less durable for heavy work. You can also sharpen plastic—if it is thick enough or hard enough—into a durable point for puncturing.

Spear Blades

To make spears, use the same procedures to make the blade that you used to make a knife blade. Then select a shaft (a straight sapling) 1.2 to 1.5 meters long. The length should allow you to handle the spear easily and effectively. Attach the spear blade to the shaft using lashing. The preferred method is to split the handle, insert the blade, then wrap or lash it tightly. You can use other materials without adding a blade. Select a 1.2- to 1.5-meter long straight hardwood shaft and shave one end to a point. If possible, fire harden the point. Bamboo also makes an excellent spear. Select a piece 1.2 to 1.5 meters long. Starting 8 to 10 centimeters back from the end used as the point, shave down the end at a 45-degree angle (Figure 2-4). Remember, to sharpen the edges, shave only the inner portion.

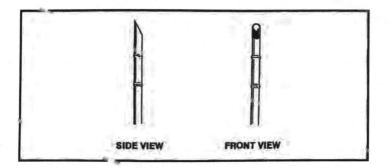

Figure 2 4; Bamboo spear

Arrow Points

To make an arrow point, use the same procedures for making a stone knife blade. Chert, flint, and shell-type stones are best for arrow points. You can fashion bone like stone—by flaking. You can make an efficient arrow point using broken glass.

OTHER EXPEDIENT WEAPONS

You can make other field-expedient weapons such as the throwing stick, archery equipment, and the bola.

Throwing Stick

The throwing stick, commonly known as the rabbit stick, is very effective against small game (squirrels, chipmunks, and rabbits). The rabbit stick itself is a blunt stick, naturally curved at about a 45-degree angle. Select a stick with the desired angle from heavy hardwood such as oak. Shave off two opposite sides so that the stick is flat like a boomerang (Figure 2-5). You must practice the throwing technique for accuracy and speed. First, align the target by extending the nonthrowing arm in line with the mid to lower section of the target. Slowly and repeatedly raise the throwing arm up and back until the throwing stick crosses the back at about a 45-degree angle c is in line with the nonthrowing hip. Bring the throwing arm forward until it is just slightly above and parallel to the nonthrowing arm This will be the throwing stick's release point. Practice slowly and repeatedly to attain accuracy.

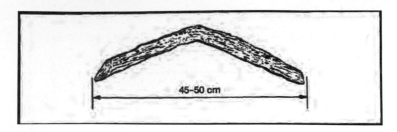

Figure 2-5: Rabbit stick

Archery Equipment

You can make a bow and arrow (Figure 2-6) from materials available in your survival area. While it may be relatively simple to make a bow and arrow, it is not easy to use one. You must practice using it a long time to be reasonably sure that you will hit your target. Also, a field-expedient bow will not last very long before you have to make a new one. For the time and effort involved, you may well decide to use another type of field-expedient weapon.

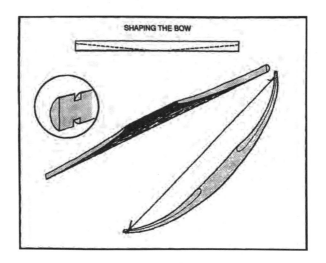

Figure 2-6: Archery equipment

Bola

The bola is another field-expedient weapon that is easy to make (Figure 2-7). It is especially effective for capturing running game or low-flying fowl in a flock. To use the bola, hold it by the center knot and twirl it above your head. Release the knot so that the bola flies toward your target. When you release the bola, the weighted cords will separate. These cords will wrap around and immobilize the fowl or animal that you hit.

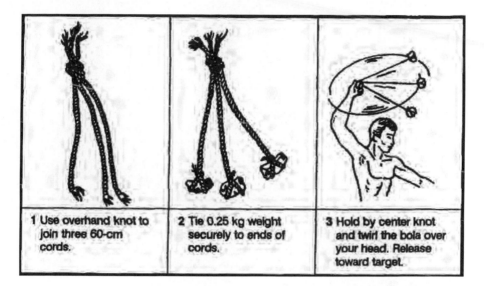

| 1 Use overhand knot to join three 60-cm cords. | 2 Tie 0.25 kg weight securely to ends of cords. | 3 Hold by center knot and twirl the bola over your head. Release toward target. |

Figure 2-7: Bola

LASHING AND CORDAGE

Many materials are strong enough for use as lashing and cordage. A number of natural and man-made materials are available in a survival situation. For example, you can make a cotton web belt much more useful by unraveling it. You can then use the string for other purposes (fishing line, thread for sewing, and lashing).

Natural Cordage Selection

Before making cordage, there are a few simple tests you can do to determine you material's suitability. First, pull on a length of the material to test for strength. Next, twist it between your fingers and roll the fibers together. If it withstands this handling and does not snap apart, tie an overhand knot with the fibers and gently tighten. If the knot does not break, the material is usable. Figure 2-8 shows various methods of making cordage.

Lashing Material

The best natural material for lashing small objects is sinew. You can make sinew from the tendons of large game, such as deer. Remove

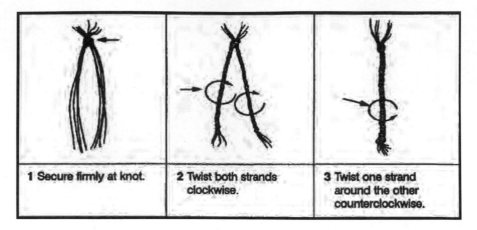

| 1 Secure firmly at knot. | 2 Twist both strands clockwise. | 3 Twist one strand around the other counterclockwise. |

Figure 2-8: Making lines from plant fibers

the tendons from the game and dry them completely. Smash the dried tendons so that they separate into fibers. Moisten the fibers and twist them into a continuous strand. If you need stronger lashing material, you can braid the strands. When you use sinew for small lashings, you do not need knots as the moistened sinew is sticky and it hardens when dry.

You can shred and braid plant fibers from the inner bark of some trees to make cord. You can use the linden, elm, hickory, white oak, mulberry, chestnut, and red and white cedar trees. After you make the cord, test it to be sure it is strong enough for your purpose. You can make these materials stronger by braiding several strands together.

You can use rawhide for larger lashing jobs. Make rawhide from the skins of medium or large game. After skinning the animal, remove any excess fat and any pieces of meat from the skin. Dry the skin completely. You do not need to stretch it as long as there are no folds to trap moisture. You do not have to remove the hair from the skin. Cut the skin while it is dry. Make cuts about 6 millimeters wide. Start from the center of the hide and make one continuous circular cut, working clockwise to the hide's outer edge. Soak the rawhide for 2 to 4 hours or until it is soft. Use it wet, stretching it as much as possible while applying it. It will be strong and durable when it dries.

RUCKSACK CONSTRUCTION

The materials for constructing a rucksack or pack are almost limitless. You can use wood, bamboo, rope, plant fiber, clothing, animal skins, canvas, and many other materials to make a pack.

There are several construction techniques for rucksacks. Many are very elaborate, but those that are simple and easy are often the most readily made in a survival situation.

Horseshoe Pack

This pack is simple to make and use and relatively comfortable to carry over one shoulder. Lay available square-shaped material, such as poncho, blanket, or canvas, flat on the ground. Lay items on one edge of the material. Pad the hard items. Roll the material (with the items) toward the opposite edge and tie both ends securely. Add extra ties along the length of the bundle. You can drape the pack over one shoulder with a line connecting the two ends (Figure 2-9).

Figure 2-9: Horseshoe pack

Square Pack

This pack is easy to construct if rope or cordage is available. Otherwise, you must first make cordage. To make this pack, construct a square frame from bamboo, limbs, or sticks. Size will vary for each person and the amount of equipment carried (Figure 2-10).

CLOTHING AND INSULATION

You can use many materials for clothing and insulation. Both man-made materials, such as parachutes, and natural materials, such as skins and plant materials, are available and offer significant protection.

Parachute Assembly

Consider the entire parachute assembly as a resource. Use every piece of material and hardware, to include the canopy, suspension lines, connector snaps, and parachute harness. Before disassembling the parachute, consider all of your survival requirements and plan

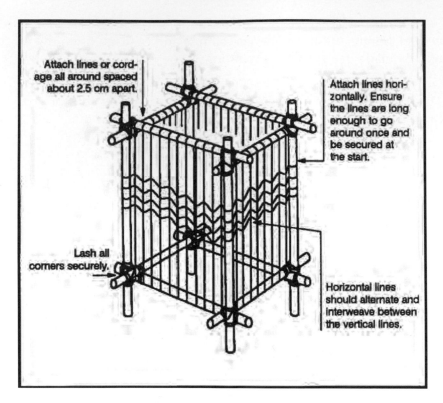

Figure 2-10: Square pack

to use different portions of the parachute accordingly. For example, consider shelter requirements, need for a rucksack, and so on, in addition to clothing or insulation needs.

Animal Skins

The selection of animal skins in a survival situation will most often be limited to what you manage to trap or hunt. However, if there is an abundance of wildlife, select the hides of larger animals with heavier coats and large fat content. Do not use the skins of infected or diseased animals if at all possible. Since they live in the wild, animals are carriers of pests such as ticks, lice, and fleas. Because of these pests, use water to thoroughly clean any skin obtained from any animal. If water is not available, at least shake out the skin thoroughly. As with rawhide, lay out the skin, and remove all fat and meat. Dry the skin completely. Use the hind quarter joint areas to make shoes and mittens or socks. Wear the hide with the fur to the inside for its insulating factor.

Plant Fibers

Several plants are sources of insulation from cold. Cattail is a marsh-land plant found along lakes, ponds, and the backwaters of rivers. The fuzz on the tops of the stalks forms dead air spaces and makes a good down-like insulation when placed between two pieces of material. Milkweed has pollen-like seeds that act as good insulation. The husk fibers from coconuts are very good for weaving ropes and, when dried, make excellent tinder and insulation.

COOKING AND EATING UTENSILS

Many materials may be used to make equipment for the cooking, eating, and storing of food.

Bowls

Use wood, bone, horn, bark, or other similar material to make bowls. To make wooden bowls, use a hollowed out piece of wood that will hold your food and enough water to cook it in. Hang the wooden container over the fire and add hot rocks to the water and food. Remove the rocks as they cool and add more hot rocks until your food is cooked.

> ### ⚠ CAUTION
> Do not use rocks with air pockets, such as limestone and sandstone. They may explode while heating in the fire.

You can also use this method with containers made of bark or leaves. However, these containers will burn above the waterline unless you keep them moist or keep the fire low.

A section of bamboo works very well, if you cut out a section between two sealed joints (Figure 2-11).

> ### ⚠ CAUTION
> A sealed section of bamboo will explode if heated because of trapped air and water in the section.

Forks, Knives, and Spoons

Carve forks, knives, and spoons from nonresinous woods so that you do not get a wood resin after taste or do not taint the food. Nonresinous woods include oak, birch, and other hardwood trees.

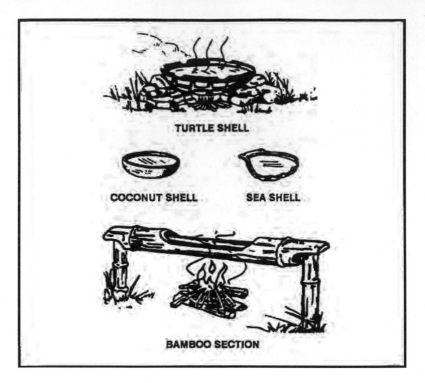

Figure 2-11: Containers for boiling food

Note: Do not use those trees that secrete a syrup or resin-like liquid on the bark or when cut.

Pots
You can make pots from turtle shells or wood. As described with bowls, using hot rocks in a hollowed out piece of wood is very effective. Bamboo is the best wood for making cooking containers.

To use turtle shells, first thoroughly boil the upper portion of the shell. Then use it to heat food and water over a flame (Figure 2-11).

Water Bottles
Make water bottles from the stomachs of larger animals. Thoroughly flush the stomach out with water, then tie off the bottom. Leave the top open, with some means of fastening it closed.

CHAPTER 3

HAND-TO-HAND COMBAT

SECTION I: OVERVIEW

Hand-to-hand combat is an engagement between two or more persons in an empty-handed struggle or with handheld weapons such as knives, sticks, and rifles with bayonets. These fighting arts are essential military skills. Projectile weapons may be lost or broken, or they may fail to fire. When friendly and enemy forces become so intermingled that firearms and grenades are not practical, hand-to-hand combat skills become vital assets.

Purpose of Combatives Training
Today's battlefield scenarios may require silent elimination of the enemy. Unarmed combat and expedient-weapons training should not be limited to forward units. With rapid mechanized/motorized, airborne, and air assault abilities, units throughout the battle area could be faced with close-quarter or unarmed fighting situations. With low-intensity conflict scenarios and guerrilla warfare conditions, any soldier is apt to face an unarmed confrontation with the enemy, and hand-to-hand combative training can save lives. The many practical battlefield benefits of combative training are not its only advantage. It can also—

    a. Contribute to individual and unit strength, flexibility, balance, and cardiorespiratory fitness.
    b. Build courage, confidence, self-discipline, and esprit de corps.

Basic Principles
There are basic principles that the hand-to-hand fighter must know and apply to successfully defeat an opponent. The principles mentioned are only a few of the basic guidelines that are essential

27

knowledge for hand-to-hand combat. There are many others, which through years of study become intuitive to a highly skilled fighter.

a. Physical Balance. Balance refers to the ability to maintain equilibrium and to remain in a stable, upright position. A hand-to-hand fighter must maintain his balance both to defend himself and to launch an effective attack. Without balance, the fighter has no stability with which to defend himself, nor does he have a base of power for an attack. The fighter must understand two aspects of balance in a struggle:

(1) How to move his body to keep or regain his own balance. A fighter develops balance through experience, but usually he keeps his feet about shoulder-width apart and his knees flexed. He lowers his center of gravity to increase stability.

(2) How to exploit weaknesses in his opponent's balance. Experience also gives the hand-to-hand fighter a sense of how to move his body in a fight to maintain his balance while exposing the enemy's weak points.

b. Mental Balance. The successful fighter must also maintain a mental balance. He must not allow fear or anger to overcome his ability to concentrate or to react instinctively in hand-to-hand combat.

c. Position. Position refers to the location of the fighter (defender) in relation to his opponent. A vital principle when being attacked is for the defender to move his body to a safe position—that is, where the attack cannot continue unless the enemy moves his whole body. To position for a counterattack, a fighter should move his whole body off the opponent's line of attack. Then, the opponent has to change his position to continue the attack. It is usually safe to move off the line of attack at a 45-degree angle, either toward the opponent or away from him, whichever is appropriate. This position affords the fighter safety and allows him to exploit weaknesses in the enemy's counterattack position. Movement to an advantageous position requires accurate timing and distance perception.

d. Timing. A fighter must be able to perceive the best time to move to an advantageous position in an attack. If he moves too soon, the enemy will anticipate his movement and adjust the attack. If the fighter moves too late, the enemy will strike him. Similarly, the fighter must launch his attack or counterattack at the critical instant when the opponent is the most vulnerable.

e. Distance. Distance is the relative distance between the positions of opponents. A fighter positions himself where distance is to his advantage. The hand-to-hand fighter must adjust his distance by changing position and developing attacks or counterattacks. He does this according to the range at which he and his opponent are engaged.

f. Momentum. Momentum is the tendency of a body in motion to continue in the direction of motion unless acted on by another force. Body mass in motion develops momentum. The greater the body mass or speed of movement, the greater the momentum. Therefore, a fighter must understand the effects of this principle and apply it to his advantage.

(1) The fighter can use his opponent's momentum to his advantage—that is, he can place the opponent in a vulnerable position by using his momentum against him.

(a) The opponent's balance can be taken away by using his own momentum.

(b) The opponent can be forced to extend farther than he expected, causing him to stop and change his direction of motion to continue his attack.

(c) An opponent's momentum can be used to add power to a fighter's own attack or counterattack by combining body masses in motion.

(2) The fighter must be aware that the enemy can also take advantage of the principle of momentum. Therefore, the fighter must avoid placing himself in an awkward or vulnerable position, and he must not allow himself to extend too far.

g. Leverage. A fighter uses leverage in hand-to-hand combat by using the natural movement of his body to place his opponent in a position of unnatural movement. The fighter uses his body or parts of his body to create a natural mechanical advantage over parts of the enemy's body. He should never oppose the enemy in a direct test of strength; however, by using leverage, he can defeat a larger or stronger opponent.

SECTION II: CLOSE-RANGE COMBATIVES

In close-range combatives, two opponents have closed the gap between them so they can grab one another in hand-to-hand combat. The principles of balance, leverage, timing, and body positioning are applied. Throws and takedown techniques are used to upset the opponent's balance and to gain control of the fight by forcing him to the ground. Chokes can be applied to quickly render an opponent unconscious. The soldier should also know counters to choking techniques to protect himself. Grappling involves skillful fighting against an opponent in close-range combat so that a soldier can win through superior body movement or grappling skills. Pain can be used to disable an opponent. A soldier can use painful eye gouges and strikes to soft, vital areas to gain an advantage over his opponent.

3-1. Throws and Takedowns

Throws and takedowns enable a hand-to-hand fighter to take an opponent to the ground where he can be controlled or disabled with further techniques. Throws and takedowns make use of the principles involved in taking the opponent's balance. The fighter uses his momentum against the attacker; he also uses leverage or body position to gain an opportunity to throw the attacker.

a. It is important for a fighter to control his opponent throughout a throw to the ground to keep the opponent from countering the throw or escaping after he is thrown to the ground. One way to do this is to control the opponent's fall so that he lands on his head. It is also imperative that a fighter maintain control of his own balance when executing throws and takedowns.

b. After executing a throw or takedown and while the opponent is on the ground, the fighter must control the opponent by any means available. He can drop his weight onto exposed areas of the opponent's body, using his elbows and knees. He can control the downed opponent's limbs by stepping on them or by placing his knees and body weight on them. Joint locks, chokes, and kicks to vital areas are also good control measures. Without endangering himself, the fighter must maintain the advantage and disable his opponent after throwing him (Figures 3-1 through 3-5).

NOTE: Although the five techniques shown in Figures 3-1 through 3-5 may be done while wearing LCE—for training purposes, it is safer to conduct all throws and takedowns without any equipment.

(1) Hip throw. The opponent throws a right punch. The defender steps in with his left foot; at the same time, he blocks the punch with his left forearm and delivers a reverse punch to the face, throat, or other vulnerable area (Figure 3-1, Step 1). (For training, deliver punches to the solar plexus.)

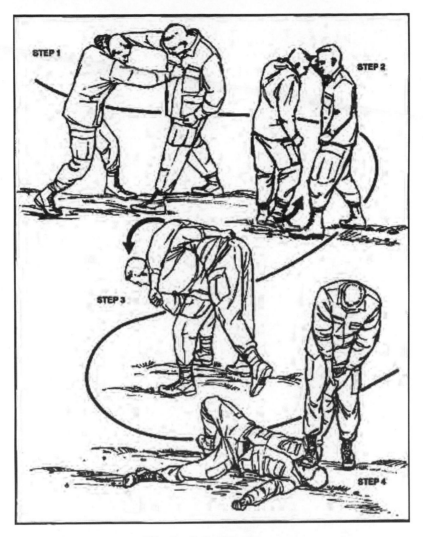

Figure 3-1: Hip throw

The defender pivots 180 degrees on the ball of his lead foot, wraps his right arm around his opponent's waist, and grasps his belt or pants (Figure 3-1, Step 2). (If opponent is wearing LCE, grasp by the pistol belt or webbing.)

The defender thrusts his hips into his opponent and maintains a grip on his opponent's right elbow. He keeps his knees shoulder-width apart and slightly bent (Figure 3-1, Step 3). He locks his knees, pulls his opponent well over his right hip, and slams him to the ground. (For training, soldier being thrown should land in a good side fall.)

By maintaining control of his opponent's arm, the defender now has the option of kicking or stomping him in the neck, face, or ribs (Figure 3-1, Step 4).

(2) Over-the-shoulder throw. The opponent lunges at the defender with a straight punch (Figure 3-2, Step 1).

The defender blocks the punch with his left forearm, pivots 180 degrees on the ball of his lead foot (Figure 3-2, Step 2), and gets well inside his opponent's right armpit with his right shoulder.

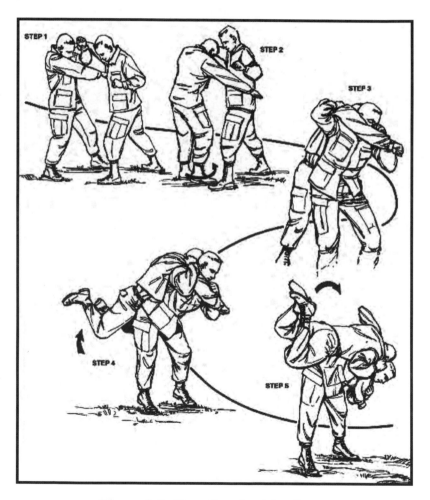

Figure 3-2: Over-the-shoulder throw

Figure 3-3: Throw from rear choke

He reaches well back under his opponent's right armpit and grasps him by the collar or hair (Figure 3-2, Step 3).

The defender maintains good back-to-chest, buttock-to-groin contact, keeping his knees slightly bent and shoulder-width apart. He maintains control of his opponent's right arm by grasping the wrist or sleeve (Figure 3-2, Step 4).

The defender bends forward at the waist and holds his opponent tightly against his body. He locks his knees, thrusts his opponent over his shoulder, and slams him to the ground (Figure 3-2, Step 5). He then has the option of disabling his opponent with kicks or stomps to vital areas.

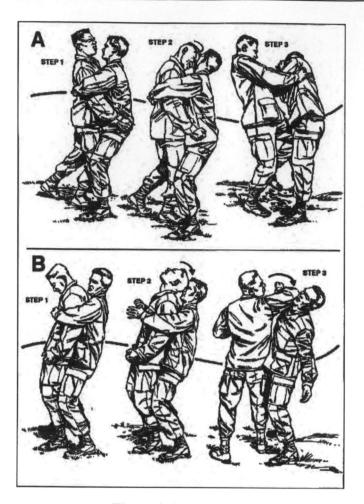

Figure 3-4: Head butt

(3) Throw from rear choke. The opponent attacks the defender with a rear strangle choke. The defender quickly bends his knees and spreads his feet shoulder-width apart (Figure 3-3, Step 1). (Knees are bent quickly to put distance between you and your opponent.)

The defender reaches as far back as possible and uses his right hand to grab his opponent by the collar or hair. He then forces his chin into the vee of the opponent's arm that is around his neck. With his left hand, he grasps the opponent's clothing at the tricep and bends forward at the waist (Figure 3-3, Step 2).

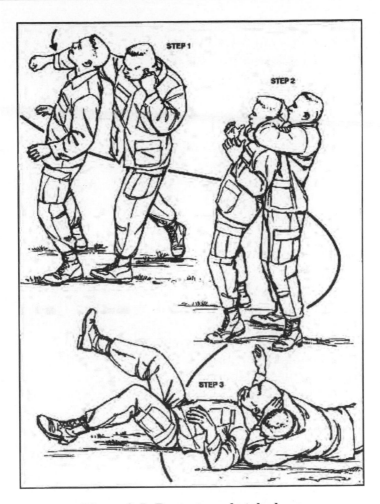

Figure 3-5: Rear strangle takedown

The defender locks his knees and, at the same time, pulls his opponent over his shoulder and slams him to the ground (Figure 3-3, Step 3).

He then has the option of spinning around and straddling his opponent or disabling him with punches to vital areas (Figure 3-3, Step 4). (It is important to grip the opponent tightly when executing this move.)

(4) Head butt. The head butt can be applied from the front or the rear. It is repeated until the opponent either releases his grip or becomes unconscious.

(a) The opponent grabs the defender in a bear hug from the front (A, Figure 3-4, Step 1).

The defender uses his forehead to smash into his opponent's nose or cheek (A, Figure 3-4, Step 2) and stuns him.

The opponent releases the defender who then follows up with a kick or knee strike to the groin (A, Figure 3-4, Step 3).

(b) The opponent grabs the defender in a bear hug from the rear (B, Figure 3-4, Step 1).

The defender cocks his head forward and smashes the back of his head into the opponent's nose or cheek area (B, Figure 3-4, Step 2).

The defender turns to face his opponent and follows up with a spinning elbow strike to the head (B, Figure 3-4, Step 3).

(5) Rear strangle takedown. The defender strikes the opponent from the rear with a forearm strike to the neck (carotid artery) (Figure 3-5, Step 1).

The defender wraps his right arm around his opponent's neck, making sure he locks the throat and windpipe in the vee formed by the his elbow. He grasps his left bicep and wraps his left hand around the back of the opponent's head. He pulls his right arm in and flexes it, pushing his opponent's head forward (Figure 3-5, Step 2).

The defender kicks his legs out and back, maintains a choke on his opponent's neck, and pulls his opponent backward until his neck breaks (Figure 3-5, Step 3).

3-2. Strangulation

Strangulation is a most effective method of disabling an opponent. The throat's vulnerability is widely known and should be a primary target in close-range fighting. Your goal may be to break the opponent's neck, to crush his trachea, to block the air supply to his lungs, or to block the blood supply to his brain.

a. Strangulation by Crushing. Crushing the trachea just below the voice box is probably one of the fastest, easiest, most lethal means of strangulation. The trachea is crushed between the thumb and first two or three fingers.

b. Respiratory Strangulation. Compressing the windpipe to obstruct air flow to the lungs is most effectively applied by pressure on the cartilage of the windpipe. Unconsciousness can take place within one to two minutes. However, the technique is not always effective on a strong opponent or an opponent with a large neck. It is better to block the blood supply to weaken the opponent first.

c. Sanguineous Strangulation. Cutting off the blood supply to the brain by applying pressure to the carotid arteries results in rapid unconsciousness of the victim. The victim can be rendered unconscious within 3 to 8 seconds, and death can result within 30 to 40 seconds.

3-3. Choking Techniques

There are several choking techniques that a soldier can use to defeat his opponent in hand-to-hand combat.

a. Cross-Collar Choke. With crossed hands, the fighter reaches as far as possible around his opponent's neck and grabs his collar (Figure 3-6, Step 1). The backs of his hands should be against the neck.

The fighter keeps his elbows bent and close to the body (as in opening a tightly sealed jar), pulls outward with both hands, and chokes the sides of the opponent's neck by rotating the knuckles into the neck (Figure 3-6, Step 2). The forearm can also be used.

b. Collar Grab Choke. The fighter grabs his opponent's collar with both hands straight-on (Figure 3-7). He then rotates the knuckles inward against the neck to quickly produce a good choke. He also keeps the elbows in front and close to the body where the greatest strength is maintained.

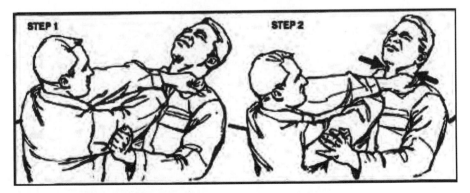

Figure 3-6: Cross-collar choke

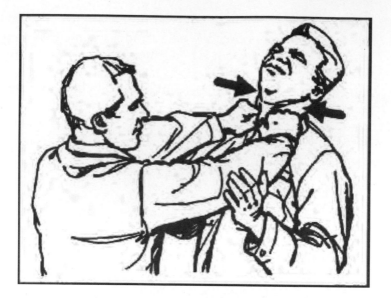

Figure 3-7: Collar grab choke

c. Carotid Choke. The fighter grabs the sides of the opponent's throat by the muscle and sticks his thumbs into the carotids, closing them off (Figure 3-8). This is a fast and painful choke.

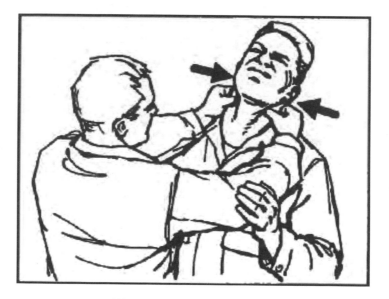

Figure 3-8: Carotid choke

d. Trachea Choke. The fighter grabs the opponent's trachea (Figure 3-9) by sticking three fingers behind the voice box on one side and the thumb behind the other. He then crushes the fingers together and twists, applying pressure until the opponent is disabled.

Figure 3-9: Trachea choke

3-4. Counters to Chokes

A soldier must know how to defend against being choked. Incapacitation and unconsciousness can occur within three seconds; therefore, it is crucial for the defender to know all possible counters to chokes.

a. Eye Gouge. The opponent attacks the defender with a frontal choke. The defender has the option of going over or under the opponent's arms. To disable the opponent, the defender inserts both thumbs into his opponent's eyes and tries to gouge them (Figure 3-10). The defender is prepared to follow-up with an attack to the vital regions.

b. Shoulder Dislocation. If the opponent applies a choke from the rear, the defender places the back of his hand against the inside of the opponent's forearm (Figure 3-11, Step 1).

Figure 3-10: Eye gouge

Then, he brings the other hand over the crook of the opponent's elbow and clasps hands, keeping his hands close to his body as he moves his entire body around the opponent (Figure 3-11, Step 2).

He positions his body so that the opponent's upper arm is aligned with the opponent's shoulders (Figure 3-11, Step 3). The opponent's arm should be bent at a 90-degree angle.

By pulling up on the opponent's elbow and down on the wrist, the opponent's balance is taken and his shoulder is easily dislocated (Figure 3-11, Step 4). The defender must use his body movement to properly position the opponent—upper body strength will not work.

He drops his body weight by bending his knees to help get the proper bend in the opponent's elbow. The defender must also keep his own hands and elbows close to his body to prevent the opponent's escape (Figure 3-11, Step 5).

c. Weight Shift. To counter being choked from above while lying on the ground (Figure 3-12, Step 1), the defender places his arms against his opponent's elbows and locks the joints.

At the same time, he shifts his hips so that his weight rests painfully on the opponent's ankle (Figure 3-12, Step 2).

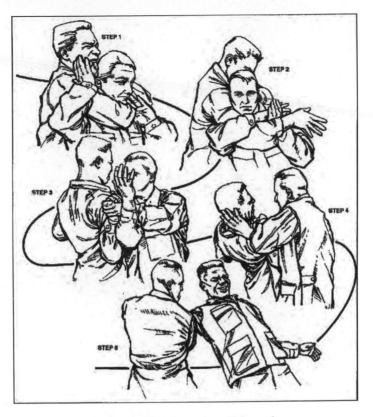

Figure 3-11: Shoulder dislocation

The defender can easily shift his body weight to gain control by turning the opponent toward his weak side (Figure 3-12, Step 3).
d. Counterstrikes to Rear Choke and Frontal Choke. As the opponent tries a rear choke (A, Figure 3- 13, Step 1), the defender can break the opponent's grip with a strong rear-elbow strike into the solar plexus (A, Figure 3-13, Step 2).
He can follow with a shin scrape down along the opponent's leg and stomp the foot (A, Figure 3-13, Step 3).
He may wish to continue by striking the groin of the opponent (A, Figure 3-13, Step 4).
As the opponent begins a frontal choke (B, Figure 3-13, Step 1), the defender turns his body and drops one arm between the opponent's arms (B, Figure 3-13, Step 2).

Figure 3-12: Weight shift

He sinks his body weight and drives his own hand to the ground, and then explodes upward with an elbow strike (B, Figure 3-13, Step 3) into the opponent's chin, stomach, or groin.

e. Headlock Escape. If a defender is in a headlock, he first turns his chin in toward his opponent's body to prevent choking (Figure 3-14, Step 1).

Next, he slides one hand up along the opponent's back, around to the face, and finds the sensitive nerve under the nose. He must avoid placing his fingers near his opponent's mouth, or he will be bitten (Figure 3-14, Step 2).

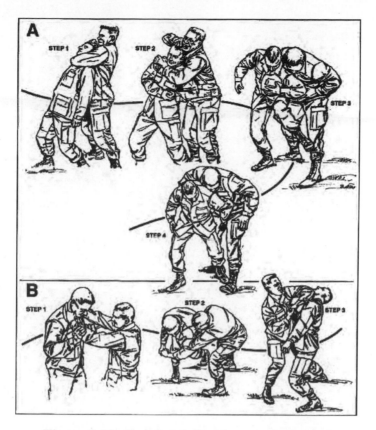

Figure 3-13: Counterstrikes to rear choke and frontal choke

The defender can now force his opponent back and then down across his own knee to the ground and maintain control by keeping pressure under the nose (Figure 3-14, Step 3). He can finish the technique with a hammer fist to the groin.

3-5. Grappling

Grappling is when two or more fighters engage in close-range, hand-to-hand combat. They may be armed or unarmed. To win, the fighter must be aware of how to move his body to maintain the upper hand, and he must know the mechanical strengths and weaknesses of the human body. The situation becomes a struggle of strength pitted against strength unless the fighter can remain in control of his opponent by using skilled movements to gain an advantage in leverage and balance. Knowledge of the following basic movement

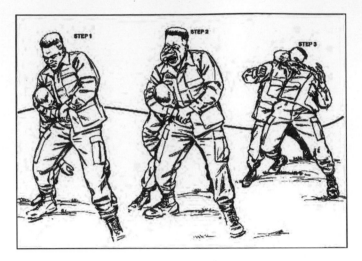

Figure 3-14: Headlock escape

techniques may give the fighter a way to apply and gain the advantage in grappling situations.

a. Wristlock From a Collar or Lapel Grab. When an opponent grabs the defender by the collar or by the lapel, the defender reaches up and grabs the opponent's hand (to prevent him from withdrawing it) while stepping back to pull him off balance (Figure 3-15, Step 1).

   The defender peels off the opponent's grabbing hand by crushing his thumb and bending it back on itself toward the palm in a straight line (Figure 3-15, Step 2). To keep his grip on the opponent's thumb, the defender keeps his hands close to his body where his control is strongest.

   He then turns his body so that he has a wristlock on his opponent. The wristlock is produced by turning his wrist outward at a 45-degree angle and by bending it toward the elbow (Figure 3-15, Step 3). The opponent can be driven to the ground by putting his palm on the ground.

b. Wristlock From an Arm Grab. When an opponent grabs a defender's arm, the defender rotates his arm to grab the opponent's forearm (Figure 3-16, Step 1).

   At the same time, he secures his other hand on the gripping hand of the opponent to prevent his escape (Figure 3-16, Step 2).

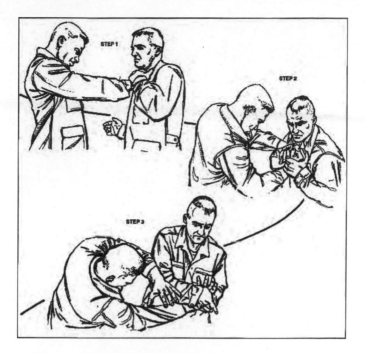

Figure 3-15: Wrist lock from collar or lapel grab

As the defender steps in toward the opponent and maintains his grip on the hand and forearm, a zee shape is formed by the opponent's arm; this is an effective wristlock (Figure 3-16, Step 3). More pain can be induced by trying to put the opponent's fingers in his own eyes.

c. Prisoner Escort. The escort secures the prisoner's arm with the wrist bent straight back upon itself, palm toward the elbow. The prisoner's elbow can be secured in the crook of the escort's elbow, firmly against the escort's body for the most control (Figure 3-17). This technique is most effective with two escorts, each holding a wrist of the prisoner. Use this technique to secure the opponent only if rope, flex cuffs, or handcuffs are unavailable.

d. Elbow Lock Against the Body. The opponent's elbow can be locked against the side of the body (Figure 3-18) by the defender. The defender turns his body to force the elbow into a position in which it was not designed to move. He can apply leverage on the opponent's wrist to gain control since the lock causes intense pain. The elbow can easily be broken to make

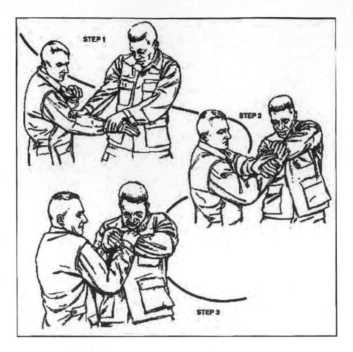

Figure 3-16: Wrist lock from arm grab

Figure 3-17: Prisoner escort

the arm ineffective. This movement must be executed with maximum speed and force.

e. Elbow Lock Against the Knee. While grappling on the ground, a defender can gain control of the situation if he can use an

Figure 3-18: Elbow lock against the body

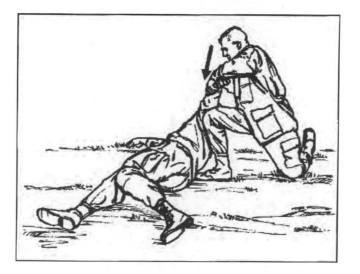

Figure 3-19: Elbow lock against the knee

elbow lock (Figure 3-19) against the opponent. He uses his knee as a fulcrum for leverage to break his opponent's arm at the elbow. Once the arm breaks, the defender must be prepared with a follow-up technique.

f. Elbow Lock Against the Shoulder. An elbow lock can be applied by locking the elbow joint against the shoulder (Figure 3-20) and pulling down on the wrist. Leverage is produced by using the shoulder as a fulcrum, by applying force, and by straightening the knees to push upward. This uses the defender's body mass and ensures more positive control. The opponent's arm must be kept straight so he cannot drive his elbow down into the defender's shoulder.

g. Shoulder Dislocation. A defender can maneuver into position to dislocate a shoulder by moving inside when an opponent launches a punch (Figure 3-21, Step 1). The defender holds his hand nearest the punching arm high to protect the head.

    The defender continues to move in and places his other arm behind the punching arm (Figure 3-21, Step 2). He strikes

Figure 3-20: Elbow lock against the shoulder

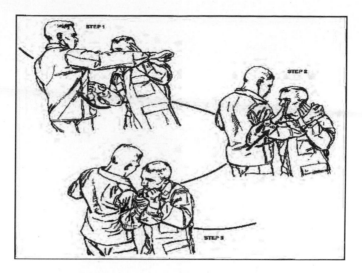

Figure 3-21: Shoulder dislocation

downward into the crook of the opponent's elbow to create a bend.

Then he clasps his hands and moves to the opponent's outside until the opponent's upper arm is in alignment with his shoulders and bent 90 degrees at the elbow. As he steps, the defender pulls up on the opponent's elbow and directs the wrist downward. This motion twists the shoulder joint so it is easily dislocated and the opponent loses his balance (Figure 3-21, Step 3).

NOTE: The defender must keep his clasped hands close to the body and properly align the opponent's arm by maneuvering his entire body. This technique will not succeed by using upperbody strength only, the opponent will escape.

(1) Straight-arm shoulder dislocation. The shoulder can also be dislocated (Figure 3-22) by keeping the elbow straight and forcing the opponent's arm backward toward the opposite shoulder at about 45 degrees. The initial movement must take the arm down and alongside the opponent's body. Bending the wrist toward the elbow helps to lock out the elbow. The dislocation also forces the opponent's head down-ward where a knee strike can be readily made. This dislocation technique

Figure 3-22: Straight-arm shoulder dislocation

should be practiced to get the feel of the correct direction in which to move the joint.

(2) Shoulder dislocation using the elbow. While grappling, the defender can snake his hand over the crook in the opponent's elbow and move his body to the outside, trapping one arm of the opponent against his side (Figure 3-23, Step 1).

The defender can then clasp his hands in front of his body and use his body mass in motion to align the opponent's upper arm with the line between the shoulders (Figure 3-23, Step 2).

By dipping his weight and then pulling upward on the opponent's elbow, the shoulder is dislocated, and the opponent loses his balance (Figure 3-23, Step 3). If the opponent's elbow locks rather than bends to allow the shoulder dislocation, the defender can use the elbow lock to keep control.

h. Knee Lock/Break. The opponent's knee joint can be attacked to produce knee locks or breaks (Figure 3-24) by forcing the knee in a direction opposite to which it was designed to move. The knee can be attacked with the body's mass behind the defender's knee or with his entire body by falling on the opponent's knee, causing it to hyperextend.

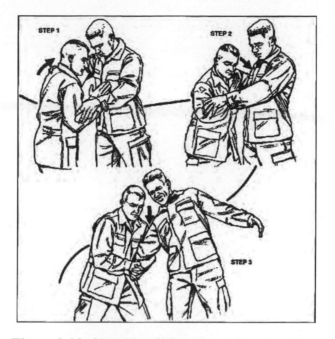

Figure 3-23: Shoulder dislocation using the elbow

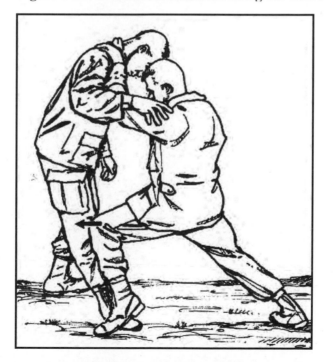

Figure 3-24: Knee lock/Break

CHAPTER 4

MEDIUM-RANGE COMBATIVES

In medium-range combatives, two opponents are already within touching distance. The arsenal of possible body weapons includes short punches and strikes with elbows, knees, and hands. Head butts are also effective; do not forget them during medium-range combat. A soldier uses his peripheral vision to evaluate the targets presented by the opponent and choose his target. He should be aggressive and concentrate his attack on the opponent's vital points to end the fight as soon as possible.

4-1. VITAL TARGETS
The body is divided into three sections: high, middle, and low. Each section contains vital targets (Figure 4-1). The effects of striking these targets follow:

    a. High Section. The high section includes the head and neck; it is the most dangerous target area.

        (1) Top of the head. The skull is weak where the frontal cranial bones join. A forceful strike causes trauma to the cranial cavity, resulting in unconsciousness and hemorrhage. A severe strike can result in death.

        (2) Forehead. A forceful blow can cause whiplash; a severe blow can cause cerebral hemorrhage and death.

        (3) Temple. The bones of the skull are weak at the temple, and an artery and large nerve lie close to the skin. A powerful strike can cause unconsciousness and brain concussion. If the artery is severed, the resulting massive hemorrhage compresses the brain, causing coma and or death.

        (4) Eyes. A slight jab in the eyes causes uncontrollable watering and blurred vision. A forceful jab or poke can cause temporary blindness, or the eyes can be gouged out. Death can result if the fingers penetrate through the thin bone behind the eyes and into the brain.

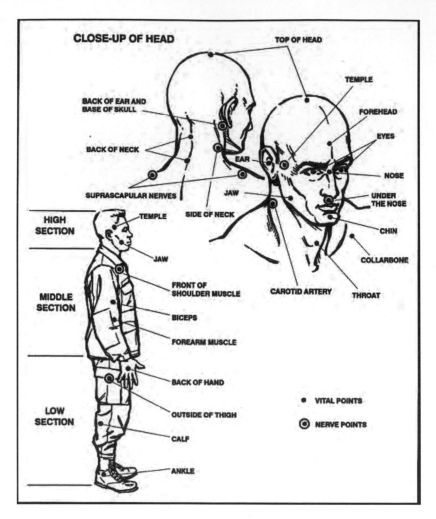

Figure 4-1A: Vital targets

(5) Ears. A strike to the ear with cupped hands can rupture the eardrum and may cause a brain concussion.

(6) Nose. Any blow can easily break the thin bones of the nose, causing extreme pain and eye watering.

(7) Under the nose. A blow to the nerve center, which is close to the surface under the nose, can cause great pain and watery eyes.

(8) Jaw. A blow to the jaw can break or dislocate it. If the facial nerve is pinched against the lower jaw, one side of the face will be paralyzed.

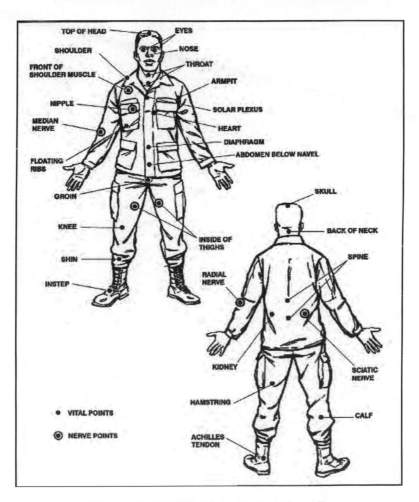

Figure 4-1B: Vital targets (continued)

(9) Chin. A blow to the chin can cause paralysis, mild concussion, and unconsciousness. The jawbone acts as a lever that can transmit the force of a blow to the back of the brain where the cardiac and respiratory mechanisms are controlled.

(10) Back of ears and base of skull. A moderate blow to the back of the ears or the base of the skull can cause unconsciousness by the jarring effect on the back of the brain. However, a powerful blow can cause a concussion or brain hemorrhage and death.

    (11) Throat. A powerful blow to the front of the throat can cause death by crushing the windpipe. A forceful blow causes extreme pain and gagging or vomiting.

    (12) Side of neck. A sharp blow to the side of the neck causes unconsciousness by shock to the carotid artery, jugular vein, and vagus nerve. For maximum effect, the blow should be focused below and slightly in front of the ear. A less powerful blow causes involuntary muscle spasms and intense pain. The side of the neck is one of the best targets to use to drop an opponent immediately or to disable him temporarily to finish him later.

    (13) Back of neck. A powerful blow to the back of one's neck can cause whiplash, concussion, or even a broken neck and death.

b. Middle Section. The middle section extends from the shoulders to the area just above the hips. Most blows to vital points in this region are not fatal but can have serious, long-term complications that range from trauma to internal organs to spinal cord injuries.

    (1) Front of shoulder muscle. A large bundle of nerves passes in front of the shoulder joint. A forceful blow causes extreme pain and can make the whole arm ineffective if the nerves are struck just right.

    (2) Collarbone. A blow to the collarbone can fracture it, causing intense pain and rendering the arm on the side of the fracture ineffective. The fracture can also sever the brachial nerve or subclavian artery.

    (3) Armpit. A large nerve lies close to the skin in each armpit. A blow to this nerve causes severe pain and partial paralysis. A knife inserted into the armpit is fatal as it severs a major artery leading from the heart.

    (4) Spine. A blow to the spinal column can sever the spinal cord, resulting in paralysis or in death.

    (5) Nipples. A large network of nerves passes near the skin at the nipples. A blow here can cause extreme pain and hemorrhage to the many blood vessels beneath.

    (6) Heart. A jolting blow to the heart can stun the opponent and allow time for follow-up or finishing techniques.

    (7) Solar plexus. The solar plexus is a center for nerves that control the cardiorespiratory system. A blow to this location is painful and can take the breath from the opponent.

A powerful blow causes unconsciousness by shock to the nerve center. A penetrating blow can also damage internal organs.

(8) Diaphragm. A blow to the lower front of the ribs can cause the diaphragm and the other muscles that control breathing to relax. This causes loss of breath and can result in unconsciousness due to respiratory failure.

(9) Floating ribs. A blow to the floating ribs can easily fracture them because they are not attached to the rib cage. Fractured ribs on the right side can cause internal injury to the liver; fractured ribs on either side can possibly puncture or collapse a lung.

(10) Kidneys. A powerful blow to the kidneys can induce shock and can possibly cause internal injury to these organs. A stab to the kidneys induces instant shock and can cause death from severe internal bleeding.

(11) Abdomen below navel. A powerful blow to the area below the navel and above the groin can cause shock, unconsciousness, and internal bleeding.

(12) Biceps. A strike to the biceps is most painful and renders the arm ineffective. The biceps is an especially good target when an opponent holds a weapon.

(13) Forearm muscle. The radial nerve, which controls much of the movement in the hand, passes over the forearm bone just below the elbow. A strike to the radial nerve renders the hand and arm ineffective. An opponent can be disarmed by a strike to the forearm; if the strike is powerful enough, he can be knocked unconscious.

(14) Back of hand. The backs of the hands are sensitive. Since the nerves pass over the bones in the hand, a strike to this area is intensely painful. The small bones on the back of the hand are easily broken and such a strike can also render the hand ineffective.

c. Low Section. The low section of the body includes everything from the groin area to the feet. Strikes to these areas are seldom fatal, but they can be incapacitating.

(1) Groin. A moderate blow to the groin can incapacitate an opponent and cause intense pain. A powerful blow can result in unconsciousness and shock.

(2) Outside of thigh. A large nerve passes near the surface on the outside of the thigh about four fingerwidths above

the knee. A powerful strike to this region can render the entire leg ineffective, causing an opponent to drop. This target is especially suitable for knee strikes and shin kicks.

(3) Inside of thigh. A large nerve passes over the bone about in the middle of the inner thigh. A blow to this area also incapacitates the leg and can cause the opponent to drop. Knee strikes and heel kicks are the weapons of choice for this target.

(4) Hamstring. A severe strike to the hamstring can cause muscle spasms and inhibit mobility. If the hamstring is cut, the leg is useless.

(5) Knee. Because the knee is a major supporting structure of the body, damage to this joint is especially detrimental to an opponent. The knee is easily dislocated when struck at an opposing angle to the joint's normal range of motion, especially when it is bearing the opponent's weight. The knee can be dislocated or hyperextended by kicks and strikes with the entire body.

(6) Calf. A powerful blow to the top of the calf causes painful muscle spasms and also inhibits mobility.

(7) Shin. A moderate blow to the shin produces great pain, especially a blow with a hard object. A powerful blow can possibly fracture the bone that supports most of the body weight.

(8) Achilles tendon. A powerful strike to the Achilles tendon on the back of the heel can cause ankle sprain and dislocation of the foot. If the tendon is torn, the opponent is incapacitated. The Achilles tendon is a good target to cut with a knife.

(9) Ankle. A blow to the ankle causes pain; if a forceful blow is delivered, the ankle can be sprained or broken.

(10) Instep. The small bones on the top of the foot are easily broken. A strike here will hinder the opponent's mobility.

4-2. STRIKING PRINCIPLES

Effective striking with the weapons of the body to the opponent's vital points is essential for a victorious outcome in a hand-to-hand struggle. A soldier must be able to employ the principles of effective striking if he is to emerge as the survivor in a fight to the death.

a. Attitude. Proper mental attitude is of primary importance in the soldier's ability to strike an opponent. In hand-to-hand combat, the soldier must have the attitude that he will defeat the enemy and complete the mission, no matter what. In a fight to the death, the soldier must have the frame of mind to survive above all else; the prospect of losing cannot enter his mind. He must commit himself to hit the opponent continuously with whatever it takes to drive him to the ground or end his resistance. A memory aid is, "Thump him and dump him!"

b. Fluid Shock Wave. A strike should be delivered so that the target is hit and the weapon remains on the impact site for at least a tenth of a second. This imparts all of the kinetic energy of the strike into the target area, producing a fluid shock wave that travels into the affected tissue and causes maximum damage. It is imperative that all strikes to vital points and nerve motor points are delivered with this principle in mind. The memory aid is, "Hit and stick!"

c. Target Selection. Strikes should be targeted at the opponent's vital points and nerve motor points. The results of effective strikes to vital points are discussed in paragraph 4-1. Strikes to nerve motor points cause temporary mental stunning and muscle motor dysfunction to the affected areas of the body. Mental stunning results when the brain is momentarily disoriented by overstimulation from too much input—for example, a strike to a major nerve. The stunning completely disables an opponent for three to seven seconds and allows the soldier to finish off the opponent, gain total control of the situation, or make his escape. Sometimes, such a strike causes unconsciousness. A successful strike to a nerve motor center also renders the affected body part immovable by causing muscle spasms and dysfunction due to nerve overload. (Readily available nerve motor points are shown in Figure 4-1.)

(1) Jugular notch pressure point. Located at the base of the neck just above the breastbone; pressure to this notch can distract and take away his balance. Pressure from fingers jabbed into the notch incurs intense pain that causes an the opponent to withdraw from the pressure involuntarily.

(2) Suprascapular nerve motor point. This nerve is located where the trapezius muscle joins the side of the neck. A strike to this point causes intense pain, temporary dysfunction of the affected arm and hand, and mental stunning for three to seven seconds. The strike should be a downward knife-hand or hammerfist strike from behind.

(3) Brachial plexus origin. This nerve motor center is on the side of the neck. It is probably the most reliable place to strike someone to stun them. Any part of the hand or arm may be applied—the palm heel, back of the hand, knife hand, ridge hand, hammer fist, thumb tip, or the forearm. A proper strike to the brachial plexus origin causes—
  • Intense pain.
  • Complete cessation of motor activity.
  • Temporary dysfunction of the affected arm.
  • Mental stunning for three to seven seconds.
  • Possible unconsciousness.

(4) Brachial plexus clavicle notch pressure point. This center is behind the collarbone in a hollow about halfway between the breastbone and the shoulder joint. The strike should be delivered with a small impact weapon or the tip of the thumb to create high-level mental stunning and dysfunction of the affected arm.

(5) Brachial plexus tie-in motor point. Located on the front of the shoulder joint, a strike to this point can cause the arm to be ineffective. Multiple strikes may be necessary to ensure total dysfunction of the arm and hand.

(6) Stellate ganglion. The ganglion is at the top of the pectoral muscle centered above the nipple. A severe strike to this center can cause high-level stunning, respiratory dysfunction, and possible unconsciousness. A straight punch or hammer fist should be used to cause spasms in the nerves affecting the heart and respiratory systems.

(7) Cervical vertebrae. Located at the base of the skull, a strike to this particular vertebrae can cause unconsciousness or possibly death. The harder the strike, the more likely death will occur.

(8) Radial nerve motor point. This nerve motor point is on top of the forearm just below the elbow. Strikes to this point can create dysfunction of the affected arm and hand. The radial nerve should be struck with the hammer fist or the forearm bones or with an impact weapon, if available. Striking the radial nerve can be especially useful when disarming an opponent armed with a knife or other weapon.

(9) Median nerve motor point. This nerve motor point is on the inside of the forearm at the base of the wrist, just above the heel of the hand. Striking this center produces similar

effects to striking the radial nerve, although it is not as accessible as the radial nerve.

(10) Sciatic nerve. A sciatic nerve is just above each buttock, but below the belt line. A substantial strike to this nerve can disable both legs and possibly cause respiratory failure. The sciatic nerve is the largest nerve in the body besides the spinal cord. Striking it can affect the entire body, especially if an impact weapon is used.

(11) Femoral nerve. This nerve is in the center of the inside of the thigh; striking the femoral nerve can cause temporary motor dysfunction of the affected leg, high-intensity pain, and mental stunning for three to seven seconds. The knee is best to use to strike the femoral nerve.

(12) Common peroneal nerve motor point. The peroneal nerve is on the outside of the thigh about four fingers above the knee. A severe strike to this center can cause collapse of the affected leg and high intensity pain, as well as mental stunning for three to seven seconds. This highly accessible point is an effective way to drop an opponent quickly. This point should be struck with a knee, shin kick, or impact weapon.

4-3. SHORT PUNCHES AND STRIKES

During medium-range combat, punches and strikes are usually short because of the close distance between fighters. Power is generated by using the entire body mass in motion behind all punches and strikes.

    a. Hands as Weapons. A knowledge of hand-to-hand combat fighting provides the fighter another means to accomplish his mission. Hands can become deadly weapons when used by a skilled fighter.

    (1) Punch to solar plexus. The defender uses this punch for close-in fighting when the opponent rushes or tries to grab him. The defender puts his full weight and force behind the punch and strikes his opponent in the solar plexus (Figure 4-2), knocking the breath out of his lungs. The defender can then follow-up with a knee to the groin, or he can use other disabling blows to vital areas.

    (2) Thumb strike to throat. The defender uses the thumb strike to the throat (Figure 4-3) as an effective technique when an opponent is rushing him or trying to grab him.

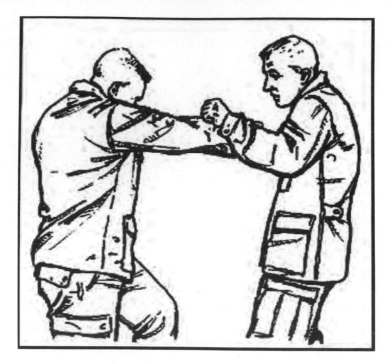

Figure 4-2: Punch to solar plexus

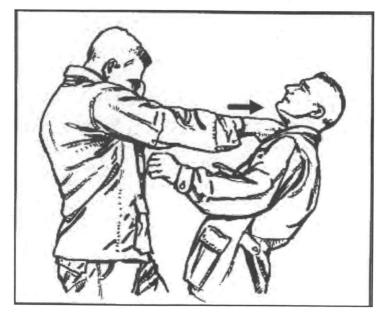

Figure 4-3: Thumb strike to throat

The defender thrusts his right arm and thumb out and strikes his opponent in the throat-larynx area while holding his left hand high for protection. He can follow up with a disabling blow to his opponent's vital areas.

(3) Thumb strike to shoulder joint. The opponent rushes the defender and tries to grab him. The defender strikes the opponent's shoulder joint or upper pectoral muscle with his fist or thumb (Figure 4-4). This technique is painful and renders the opponent's arm numb. The defender then follows up with a disabling movement.

Figure 4-4: Thumb strike to shoulder joint

(4) Hammer-fist strike to face. The opponent rushes the defender. The defender counters by rotating his body in the direction of his opponent and by striking him in the temple, ear, or face (Figure 4-5). The defender follows up with kicks to the groin or hand strikes to his opponent's other vital areas.

(5) Hammer-fist strike to side of neck. The defender catches his opponent off guard, rotates at the waist to generate power, and strikes his opponent on the side of the neck (carotid artery) (Figure 4-6) with his hand clenched into a fist. This strike can cause muscle spasms at the least and may knock his opponent unconscious.

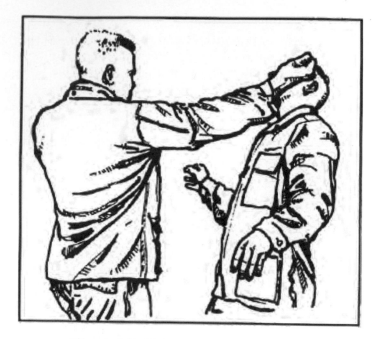

Figure 4-5: Hammer-fist strike to face

Figure 4-6: Hammer-fist strike to neck

(6) Hammer fist to pectoral muscle. When the opponent tries to grapple with the defender, the defender counters by forcefully striking his opponent in the pectoral muscle (Figure 4-7). This blow stuns the opponent, and the defender immediately follows up with a disabling blow to a vital area of his opponent's body.

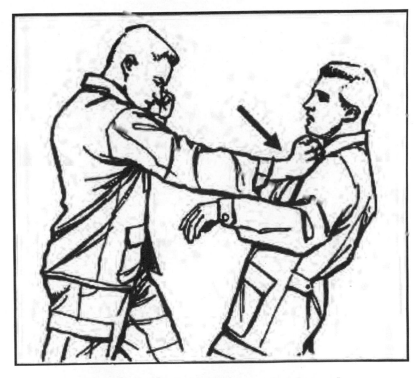

Figure 4-7: Hammer-fist to pectoral muscle

(7) Hook punch to solar plexus or floating ribs. The opponent tries to wrestle the defender to the ground. The defender counters with a short hook punch to his opponent's solar plexus or floating ribs (Figure 4-8). A sharply delivered blow can puncture or collapse a lung. The defender then follows up with a combination of blows to his opponent's vital areas.

(8) Uppercut to chin. The defender steps between his opponent's arms and strikes with an uppercut punch (Figure 4-9) to the chin or jaw. The defender then follows up with blows to his opponent's vital areas.

Figure 4-8: Hook punch to solar plexus or floating ribs

Figure 4-9: Uppercut to chin

(9) Knife-hand strike to side of neck. The defender executes a knife-hand strike to the side of his opponent's neck (Figure 4-10) the same way as the hammer-fist strike (Figure 4-6) except he uses the edge of his striking hand.

Figure 4-10: Knife-hand to side of neck

(10) Knife-hand strike to radial nerve. The opponent tries to strike the defender with a punch. The defender counters by striking his opponent on the top of the forearm just below the elbow (radial nerve) (Figure 4-11) and uses a follow-up technique to disable his opponent.

(11) Palm-heel strike to chin. The opponent tries to surprise the defender by lunging at him. The defender quickly counters by striking his opponent with a palm-heel strike to the chin (Figure 4- 12), using maximum force.

(12) Palm-heel strike to solar plexus. The defender meets his opponent's rush by striking him with a palm-heel strike to the solar plexus (Figure 4-13). The defender then executes a follow-up technique to his opponent's vital organs.

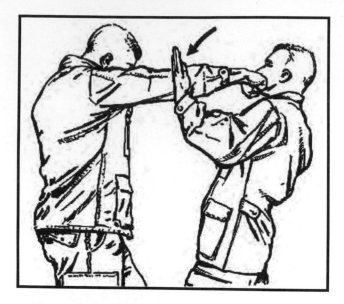

Figure 4-11: Knife-hand strike to radial nerve

Figure 4-12: Palm heel strike to chin

Figure 4-13: Palm-heel strike to solar plexus

(13) Palm-heel strike to kidneys. The defender grasps his opponent from behind by the collar and pulls him off balance. He quickly follows up with a hard palm-heel strike to the opponent's kidney (Figure 4-14). The defender can then take down his opponent with a follow-up technique to the back of his knee.

b. Elbows as Weapons. The elbows are also formidable weapons; tremendous striking power can be generated from them. The point of the elbow should be the point of impact. The elbows are strongest when kept in front of the body and in alignment with the shoulder joint; that is, never strike with the elbow out to the side of the body.

(1) Elbow strikes. When properly executed, elbow strikes (Figures 4-15 through 4-21) render an opponent ineffective. When using elbow strikes, execute them quickly, powerfully, and repetitively until the opponent is disabled.

Figure 4-14: Palm-heel strike to kidneys

Figure 4-15: Elbow strike to face

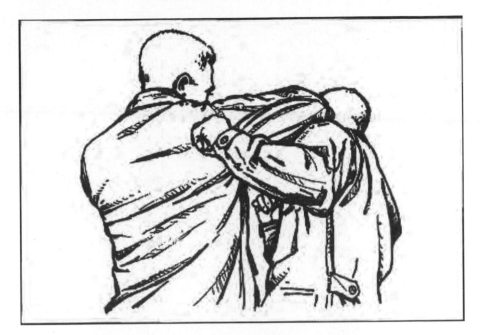

Figure 4-16: Elbow strike to temple

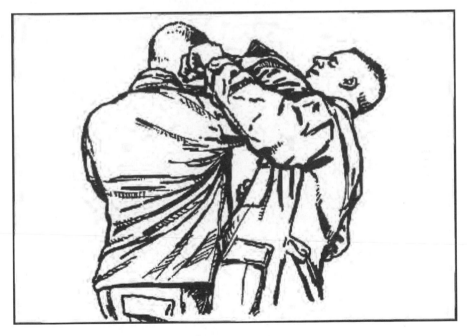

Figure 4-17: Rising elbow strike

Figure 4-18: Elbow strike to head

Figure 4-19: Elbow strike to solar plexus

Figure 4-20: Elbow strike to biceps

Figure 4-21: Elbow strike to inside of shoulder

(2) Repetitive elbow strikes. The attacker on the right throws a punch (Figure 4-22, Step 1).

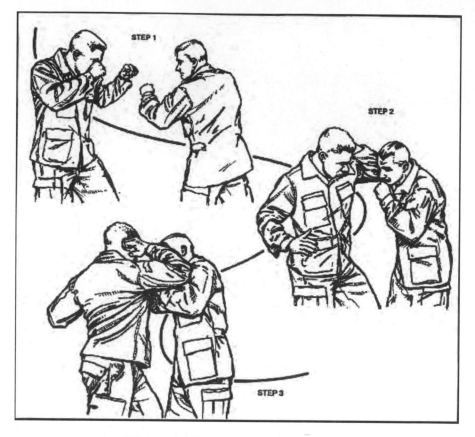

Figure 4-22: Repetitive elbow strike

The defender counters with an elbow strike to the biceps (Figure 4-22, Step 2). The attacker follows with a punch from his other arm.

The defender again counters with an elbow strike to the shoulder joint (Figure 4-22, Step 3). He next strikes with an elbow from the opposite side to the throat.

c. Knees as Weapons. When the knees are used to strike opponents, they are especially potent weapons and are hard to defend or protect against. Great power is generated by thrusting the hips in with a knee strike; however, use the point of the knee as the impact surface. All knee strikes should be executed

repetitively until the opponent is disabled. The following techniques are the most effective way to overpower or disable the
opponent.

(1) Front knee strike. When an opponent tries to grapple with
the defender, the defender strikes his opponent in the
stomach or solar plexus with his knee (Figure 4-23). This
stuns the opponent and the defender can follow up with
another technique.

Figure 4-23: Front knee strike

(2) Knee strike to outside of thigh. The defender delivers a
knee strike to the outside of his opponent's thigh (common
peroneal nerve) (Figure 4-24). This strike causes intense
pain and renders the opponent's leg ineffective.

Figure 4-24: Knee to outside of thigh

(3) Knee strike to inside of thigh. An effective technique for close-in grappling is when the defender delivers a knee strike to the inside of his opponent's thigh (peroneal nerve) (Figure 4-25). The defender then executes a follow-up technique to a vital point.

Figure 4-25: Knee to inside of thigh

(4) Knee strike to groin. The knee strike to the groin is effective during close-in grappling. The defender gains control by grabbing his opponent's head, hair, ears, or shoulders and strikes him in the groin with his knee (Figure 4-26).

Figure 4-26: Knee strike to groin

(5) Knee strike to face. The defender controls his opponent by grabbing behind his head with both hands and forcefully pushing his head down. At the same time, the defender brings his knee up and smashes the opponent in the face (Figure 4-27). When properly executed, the knee strike to the face is a devastating technique that can cause serious injury to the opponent.

Figure 4-27: Knee strike to face

CHAPTER 5

LONG-RANGE COMBATIVES

In long-range combatives, the distance between opponents is such that the combatants can engage one another with fully extended punches and kicks or with handheld weapons, such as rifles with fixed bayonets and clubs. As in medium-range combatives, a fighter must continuously monitor his available body weapons and opportunities for attack, as well as possible defense measures. He must know when to increase the distance from an opponent and when to close the gap. The spheres of influence that surround each fighter come into contact in long-range combatives. (See Chapter 6 for interval gaps and spheres of influence.)

SECTION I: NATURAL WEAPONS

The most dangerous natural weapons a soldier possesses are his hands and feet. This section describes natural weapon techniques of various punches, strikes, and kicks and stresses aggressive tactics with which to subdue an opponent.

5-1. Extended Arm Punches and Strikes

Extended arm punches and strikes in long-range combatives, like those in medium-range combatives, should be directed at vital points and nerve motor points. It is essential to put the entire body mass in motion behind long-range strikes. Closing the distance to the target gives the fighter an opportunity to take advantage of this principle.

    a. In extended punches, the body weapon is usually the fist, although the fingers may be used—for example, eye gouging. When punching, hold the fist vertically or horizontally. Keep the wrist straight to prevent injury and use the first two knuckles in striking.

    b. Another useful variation of the fist is to place the thumb on top of the vertical fist so that the tip protrudes beyond the curled index finger that supports it. The thumb strike is especially effective against soft targets. Do not fully lock out the arm when punching; keep a slight bend in the elbow to prevent hyperextension if the intended target is missed.

5-2. Kicks

Kicks during hand-to-hand combat are best directed to low targets and should be simple but effective. Combat soldiers are usually burdened with combat boots and LCE. His flexibility level is usually low during combat, and if engaged in hand-to-hand combat, he will be under high stress. He must rely on gross motor skills and kicks that do not require complicated movement or much training and practice to execute.

a. Side Knee Kick. When an opponent launches an attack—for example, with a knife (Figure 5-1, Step 1), it is most important for the defender to first move his entire body off the line of attack as the attacker moves in.

Figure 5-1: Side knee kick

As the defender steps off at 45 degrees to the outside and toward the opponent, he strikes with a short punch to the floating ribs (Figure 5-1, Step 2).

Then the defender turns his body by rotating on the leading, outside foot and raises the knee of his kicking leg to his chest. He then drives his kick into the side of the attacker's knee with his foot turned 45 degrees outward (Figure 5-1, Step 3). This angle makes the most of the striking surface and reduces his chances of missing the target.

b. Front Knee Kick. As the attacker moves in, the defender immediately shifts off the line of attack and drives his kicking foot straight into the knee of the attacker (Figure 5-2). He turns his foot 45 degrees to make the most of the striking surface and to reduce the chances of missing the target. If the kick is done right, the attacker's advance will stop abruptly, and the knee joint will break.

c. Heel Kick to Inside of Thigh. The defender steps 45 degrees outside and toward the attacker to get off the line of attack. He is now in a position where he can drive his heel into the inside of the opponent's thigh (femoral nerve) (Figure 5-3, Steps 1 and 2). Either thigh can be targeted because the kick can still be executed if the defender moves to the inside of the

Figure 5-2: Front knee kick

Figure 5-3: Heel kick to inside of thigh

opponent rather than to the outside when getting off the line
of attack.

d. Heel Kick to Groin. The defender drives a heel kick into the
attacker's groin (Figure 5-4) with his full body mass behind it.
Since the groin is a soft target, the toe can also be used when
striking it.

e. Shin Kick. The shin kick is a powerful kick, and it is easily
performed with little training. When the legs are targeted, the
kick is hard to defend against (Figure 5-5), and an opponent
can be dropped by it.

The calves and common peroneal nerve (Figure 5-6) are the
best striking points.

The shin kick can also be used to attack the floating ribs
(Figure 5-7).

f. Stepping Side Kick. A soldier starts a stepping side kick (Figure 5-8, Step 1) by stepping either behind or in front of his
other foot to close the distance between him and his opponent.
The movement is like that in a skip.

Figure 5-4: Heel kick to groin

Figure 5-5: Shin kick to legs

Figure 5-6: Shin kick to common peroneal nerve

Figure 5-7: Shin kick to floating ribs

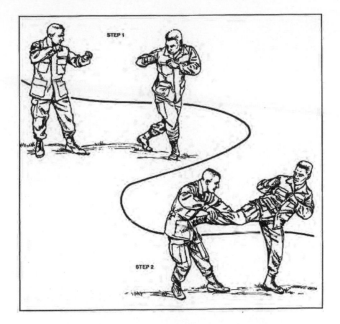

Figure 5-8: Stepping side kick

The soldier now brings the knee of his kicking foot up and thrusts out a side kick (Figure 5-8, Step 2). Tremendous power and momentum can be developed in this kick.

g. Counter to Front Kick. When the attacker tries a front kick, the defender traps the kicking foot by meeting it with his own (Figure 5-9, Step 1). The defender turns his foot 45 degrees outward to increase the likelihood of striking the opponent's kicking foot. This counter requires good timing by the defender, but not necessarily speed. Do not look at the feet; use your peripheral vision.

When an attacker tries a front kick (Figure 5-9, Step 2), the defender steps off the line of attack of the incoming foot to the outside.

As the attacker's kicking leg begins to drop, the defender kicks upward into the calf of the attacker's leg (Figure 5-9, Step 3). This kick is extremely painful and will probably render the leg ineffective. This technique does not rely on the defender's speed, but on proper timing.

The defender can also kick to an opponent's kicking leg by moving off the line of attack to the inside and by using

Figure 5-9: Counter to front kick

the heel kick to the inside of the thigh or groin (Figure 5-9, Step 4).

h. Counter to Roundhouse-Type Kick. When an opponent prepares to attack with a roundhouse-type kick (Figure 5-10, Step 1), the defender moves off the line of attack by stepping to the inside of the knee of the kicking leg.

He then turns his body to receive the momentum of the leg (Figure 5-10, Step 2). By moving to the inside of the knee, the defender lessens the power of the attacker's kicking leg. The harder the attacker kicks, the more likely he is to hyperextend his own knee against the body of the defender, but the defender will not be harmed. However, the defender must get to the inside of the knee, or an experienced opponent can change his roundhouse kick into a knee strike. The defender receives the energy of the kicking leg and continues turning with the momentum of the kick.

Figure 5-10: Counter to roadhouse kick

The attacker will be taken down by the defender's other leg with no effort (Figure 5-10, Step 3).

i. Kick as a Defense Against Punch. As the opponent on the left throws a punch (Figure 5-11, Step 1), the defender steps off the line of attack to the outside.

He then turns toward the opponent, brings his knee to his chest, and launches a heel kick to the outside of the opponent's thigh (Figure 5-11, Step 2). He keeps his foot turned 45 degrees to ensure striking the target and to maintain balance.

SECTION II: DEFENSIVE TECHNIQUES

A knife (or bayonet), properly employed, is a deadly weapon; however, using defensive techniques, such as maintaining separation, will greatly enhance the soldier's ability to fight and win.

5-3. Defense Against an Armed Opponent

An unarmed defender is always at a distinct disadvantage facing an armed opponent. It is imperative therefore that the unarmed defender understand and use the following principles to survive:

a. Separation. Maintain a separation of at least 10 feet plus the length of the weapon from the attacker. This distance gives the defender time to react to any attempt by the attacker to close the gap and be upon the defender. The defender should also try to place stationary objects between himself and the attacker.

b. Unarmed Defense. Unarmed defense against an armed opponent should be a last resort. If it is necessary, the defender's course of action includes:

   (1) Move the body out of the line of attack of the weapon. Step off the line of attack or redirect the attack of the weapon so that it clears the body.

   (2) Control the weapon. Maintain control of the attacking arm by securing the weapon, hand, wrist, elbow, or arm by using joint locks, if possible.

Figure 5-11: Kick as a defense against punch

(3) Stun the attacker with an effective counterattack. Counterattack should be swift and devastating. Take the vigor out of the attacker with a low, unexpected kick, or break a locked joint of the attacking arm. Strikes to motor nerve centers are effective stuns, as are skin tearing, eye gouging, and attacking of the throat. The defender can also take away the attacker's balance.

(4) Ground the attacker. Take the attacker to the ground where the defender can continue to disarm or further disable him.

(5) Disarm the attacker. Break the attacker's locked joints. Use leverage or induce pain to disarm the attacker and finish him or to maintain physical control.

c. Precaution. Do not focus full attention on the weapon because the attacker has other body weapons to use. There may even be other attackers that you have not seen.

d. Expedient Aids. Anything available can become an expedient aid to defend against an armed attack. The kevlar helmet can be used as a shield; similarly, the LCE and shirt jacket can be used to protect the defender against a weapon. The defender can also throw dirt in the attacker's eyes as a distraction.

5-4. Angles of Attack
Any attack, regardless of the type weapon, can be directed along one of nine angles (Figure 5-12). The defense must be oriented for each angle of attack.

a. No. 1 Angle of Attack. A downward diagonal slash, stab, or strike toward the left side of the defender's head, neck, or torso.

b. No. 2 Angle of Attack. A downward diagonal slash, stab, or strike toward the right side of the defender's head, neck, or torso.

c. No. 3 Angle of Attack. A horizontal attack to the left side of the defender's torso in the ribs, side, or hip region.

d. No. 4 Angle of Attack. The same as No. 3 angle, but to the right side.

e. No. 5 Angle of Attack. A jabbing, lunging, or punching attack directed straight toward the defender's front.

f. No. 6 Angle of Attack. An attack directed straight down upon the defender.

g. No. 7 Angle of Attack. An upward diagonal attack toward the defender's lower-left side.

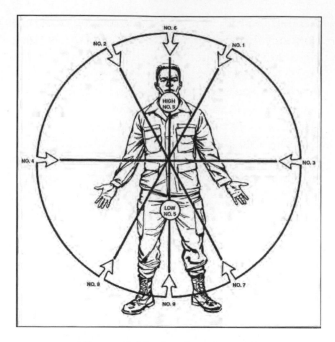

Figure 5-12: Angles of attack

    h. No. 8 Angle of Attack. An upward diagonal attack toward the defender's lower-right side.

    i. No. 9 Angle of Attack. An attack directed straight up—for example, to the defender's groin.

5-5. Defense Against a Knife

When an unarmed soldier is faced with an enemy armed with a knife, he must be mentally prepared to be cut. The likelihood of being cut severely is less if the fighter is well trained in knife defense and if the principles of weapon defense are followed. A slash wound is not usually lethal or shock inducing; however, a stab wound risks injury to vital organs, arteries, and veins and may also cause instant shock or unconsciousness.

    a. Types of Knife Attacks. The first line of defense against an opponent armed with a knife is to avoid close contact. The different types of knife attacks follow:

        (1) Thrust. The thrust is the most common and most dangerous type of knife attack. It is a strike directed straight into the target by jabbing or lunging.

(2) Slash. The slash is a sweeping surface cut or circular slash. The wound is usually a long cut, varying from a slight surface cut to a deep gash.

(3) Flick. This attack is delivered by flicking the wrist and knife to extended limbs, inflicting numerous cuts. The flick is very distractive to the defender since he is bleeding from several cuts if the attacker is successful.

(4) Tear. The tear is a cut made by dragging the tip of the blade across the body to create a ripping type cut.

(5) Hack. The hack is delivered by using the knife to block or chop with.

(6) Butt. The butt is a strike with the knife handle.

b. Knife Defense Drills. Knife defense drills are used to familiarize soldiers with defense movement techniques for various angles of attack. For training, the soldiers should be paired off; one partner is named as the attacker and one is the defender. It is important that the attacker make his attack realistic in terms of distance and angling during training. His strikes must be accurate in hitting the defender at the intended target if the defender does not defend himself or move off the line of attack. For safety, the attacks are delivered first at one-quarter and one-half speed, and then at three-quarter speed as the defender becomes more skilled. Variations can be added by changing grips, stances, and attacks.

(1) No. 1 angle of defense—check and lift. The attacker delivers a slash along the No. 1 angle of attack. The defender meets and checks the movement with his left forearm bone, striking the inside forearm of the attacker (Figure 5-13, Step 1).

The defender's right hand immediately follows behind the strike to lift, redirect, and take control of the attacker's knife arm (Figure 5-13, Step 2).

The defender brings the attacking arm around to his right side where he can use an arm bar, wrist lock, and so forth, to disarm the attacker (Figure 5-13, Step 3).

He will have better control by keeping the knife hand as close to his body as possible (Figure 5-13, Step 4).

(2) No. 2 angle of defense—check and ride. The attacker slashes with a No. 2 angle of attack. The defender meets the attacking arm with a strike from both forearms against the outside forearm, his bone against the attacker's muscle tissue (Figure 5-14, Step 1).

**Figure 5-13: No. 1 angle of defense—
check and lift**

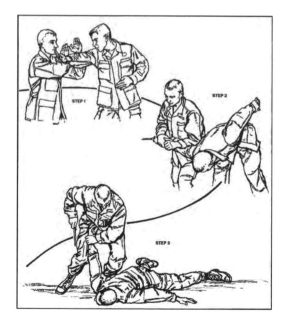

**Figure 5-14: No. 2 angle of defense—
check and ride**

The strike checks the forward momentum of the attacking arm. The defender's right hand is then used to ride the attacking arm clear of his body (Figure 5-14, Step 2).

He redirects the attacker's energy with strength starting from the right elbow (Figure 5-14, Step 3).

(3) No. 3 angle of defense—check and lift. The attacker delivers a horizontal slash to the defender's ribs, kidneys, or hip on the left side (Figure 5-15, Step 1). The defender meets and checks the attacking arm on the left side of his body with a downward circular motion across the front of his own body.

At the same time, he moves his body off the line of attack. He should meet the attacker's forearm with a strike forceful enough to check its momentum (Figure 5-15, Step 2). The defender then rides the energy of the attacking arm by wiping downward along the outside of his own left forearm with his right hand.

He then redirects the knife hand around to his right side where he can control or disarm the weapon (Figure 5-15, Step 3).

**Figure 5-15: No. 3 angle of defense—
check and lift**

(4) No. 4 angle of defense—check. The attacker slashes the defender with a backhand slashing motion to the right side at the ribs, kidneys, or hips. The defender moves his right arm in a downward circular motion and strikes the attacking arm on the outside of the body (Figure 5-16, Step 1).

At the same time, he moves off the line of attack (Figure 5-16, Step 2). The strike must be forceful enough to check the attack.

The left arm is held in a higher guard position to protect from a redirected attack or to assist in checking (Figure 5-16, Step 3).

The defender moves his body to a position where he can choose a proper disarming maneuver (Figure 5-16, Step 4).

(5) Low No. 5 angle of defense—parry. A lunging thrust to the stomach is made by the attacker along the No. 5 angle of attack (Figure 5-17, Step 1).

The defender moves his body off the line of attack and deflects the attacking arm by parrying with his left hand (Figure 5-17, Step 2). He deflects the attacking hand toward his right side by redirecting it with his right hand.

Figure 5-16: No. 4 angle of defense--check

As he does this, the defender can strike downward with the left forearm or the wrist onto the forearm or wrist of the attacker (Figure 5-17, Step 3).

The defender ends up in a position to lock the elbow of the attacking arm across his body if he steps off the line of attack properly (Figure 5-17, Step 4).

(6) High No. 5 angle of defense. The attacker lunges with a thrust to the face, throat, or solar plexus (Figure 5-18, Step 1).

The defender moves his body off the line of attack while parrying with either hand. He redirects the attacking arm so that the knife clears his body (Figure 5-18, Step 2).

He maintains control of the weapon hand or arm and gouges the eyes of the attacker, driving him backward and off balance (Figure 5-18, Step 3). If the attacker is much taller than the defender, it may be a more natural movement for the defender to raise his left hand to strike and deflect the attacking arm. He can then gouge his thumb or

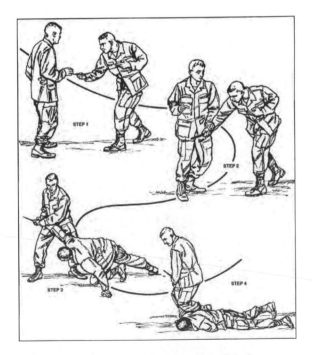

**Figure 5-17: Low No. 5 angle of defense--
parry**

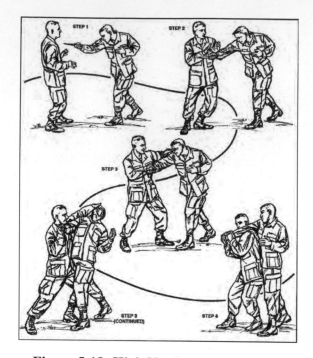

Figure 5-18: High No. 5 angle of defense

fingers into the jugular notch of the attacker and force him to the ground.

Still another possibility for a high No. 5 angle of attack is for the defender to move his body off the line of attack while parrying. He can then turn his body, rotate his shoulder under the elbow joint of the attacker, and lock it out (Figure 5-18, Step 4).

(7) No. 6 angle of defense. The attacker strikes straight downward onto the defender with a stab (Figure 5-19, Step 1).

The defender reacts by moving his body out of the weapon's path and by parrying or checking and redirecting the attacking arm, as the movement in the high No. 5 angle of defense (Figure 5-19, Step 2). The reactions may vary as to what is natural for the defender.

The defender then takes control of the weapon and disarms the attacker (Figure 5-19, Step 3).

c. Follow-Up Techniques. Once the instructor believes the soldiers are skilled in these basic reactions to attack, follow-up techniques may be introduced and practiced. These drills

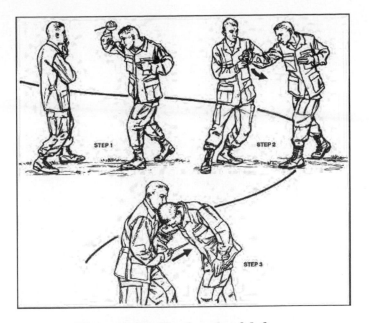

Figure 5-19: No. 6 angle of defense

make up the defense possibilities against the various angles of attack. They also enable the soldier to apply the principles of defense against weapons and allow him to feel the movements. Through repetition, the reactions become natural, and the soldier instinctively reacts to a knife attack with the proper defense. It is important not to associate specific movements or techniques with certain types of attack. The knife fighter must rely on his knowledge of principles and his training experience in reacting to a knife attack. No two attacks or reactions will be the same; thus, memorizing techniques will not ensure a soldier's survival.

(1) Defend and clear. When the defender has performed a defensive maneuver and avoided an attack, he can push the attacker away and move out of the attacker's reach.

(2) Defend and stun. After the defender performs his first defensive maneuver to a safer position, he can deliver a stunning blow as an immediate counterattack. Strikes to motor nerve points or attacker's limbs, low kicks, and elbow strikes are especially effective stunning techniques.

(3) Defend and disarm. The defender also follows up his first defensive maneuver by maintaining control of the attacker's weapon arm, executing a stunning technique, and disarming the attacker. The stun distracts the attacker and also gives the defender some time to gain possession of the weapon and to execute his disarming technique.

5-6. Unarmed Defense Against a Rifle with Fixed Bayonet

Defense against a rifle with a fixed bayonet involves the same principles as knife defense. The soldier considers the same angles of attack and the proper response for any attack along each angle.

a. Regardless of the type weapon used by the enemy, his attack will always be along one of the nine angles of attack at any one time. The soldier must get his entire body off the line of attack by moving to a safe position. A rifle with a fixed bayonet has two weapons: a knife at one end and a butt stock at the other end. The soldier will be safe as long as he is not in a position where he can be struck by either end during the attack.

b. Usually, he is in a more advantageous position if he moves inside the length of the weapon. He can then counterattack to gain control of the situation as soon as possible. The following counterattacks can be used as defenses against a rifle with a fixed bayonet; they also provide a good basis for training.

(1) Unarmed defense against No. 1 angle of attack. The attacker prepares to slash along the No. 1 angle of attack (Figure 5-20, Step 1).

The defender waits until the last possible moment before moving so he is certain of the angle along which the attack is directed (Figure 5-20, Step 2). This way, the attacker cannot change his attack in response to movement by the defender.

When the defender is certain that the attack is committed along a specific angle (No. 1, in this case), he moves to the inside of the attacker and gouges his eyes (Figure 5-20, Step 2) while the other hand redirects and controls the weapon. He maintains control of the weapon and lunges his entire body weight into the eye gouge to drive the attacker backward and off balance. The defender now ends up with the weapon, and the attacker is in a poor recovery position (Figure 5-20, Step 3).

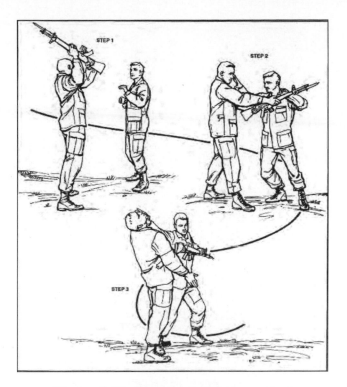

**Figure 5-20: Unarmed defense against
No. 1 angle of attack**

(2) Unarmed defense against No. 2 angle of attack. The attacker makes a diagonal slash along the No. 2 angle of attack (Figure 5-21, Step 1). Again, the defender waits until he is sure of the attack before moving.

  The defender then moves to the outside of the attacker and counterattacks with a thumb jab into the right armpit (Figure 5-21, Step 2). He receives the momentum of the attacking weapon and controls it with his free hand.

  He uses the attacker's momentum against him by pulling the weapon in the direction it is going with one hand and pushing with his thumb of the other hand (Figure 5-21, Step 3). The attacker is completely off balance, and the defender can gain control of the weapon.

(3) Unarmed defense against No. 3 angle of attack. The attacker directs a horizontal slash along the No. 3 angle of attack (Figure 5-22, Step 1).

**Figure 5-21: Unarmed defense against No. 2
angle of attack**

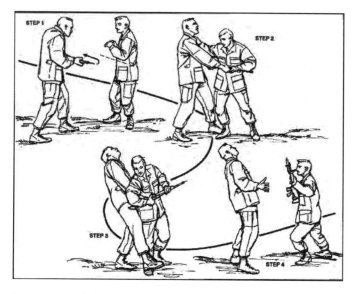

**Figure 5-22: Unarmed defense against No. 3 angle of
attack**

The defender turns and moves to the inside of the attacker; he then strikes with his thumb into the jugular notch (Figure 5-22, Step 2).

His entire body mass is behind the thumb strike and, coupled with the incoming momentum of the attacker, the strike drives the attacker's head backward and takes his balance (Figure 5-22, Step 3).

The defender turns his body with the momentum of the weapon's attack to strip the weapon from the attacker's grip (Figure 5-22, Step 4).

(4) Unarmed defense against No. 4 angle of attack. The attack is a horizontal slash along the No. 4 angle of attack (Figure 5-23, Step 1).

The defender moves into the outside of the attacker (Figure 5-23, Step 2).

He then turns with the attack, delivering an elbow strike to the throat (Figure 5-23, Step 3). At the same time, the defender's free hand controls the weapon and pulls it from the attacker as he is knocked off balance from the elbow strike.

**Figure 5-23: Unarmed defense against No. 4 angle
of attack**

(5) Unarmed defense against low No. 5 angle of attack. The attacker thrusts the bayonet at the stomach of the defender (Figure 5-24, Step 1).

The defender shifts his body to the side to avoid the attack and to gouge the eyes of the attacker (Figure 5-24, Step 2).

The defender's free hand maintains control of and strips the weapon from the attacker as he is driven backward with the eye gouge (Figure 5-24, Step 3).

(6) Unarmed defense against high No. 5 angle of attack. The attacker delivers a thrust to the throat of the defender (Figure 5-25, Step 1).

The defender then shifts to the side to avoid the attack, parries the thrust, and controls the weapon with his trail hand (Figure 5-25, Step 2).

He then shifts his entire body mass forward over the lead foot, slamming a forearm strike into the attacker's throat (Figure 5-25, Step 3).

(7) Unarmed defense against No 6 angle of attack. The attacker delivers a downward stroke along the No. 6 angle of attack (Figure 5-26, Step 1).

**Figure 5-24: Unarmed defense against low No. 5
angle of attack**

Figure 5-25: Unarmed defense against high No. 5 angle of attack

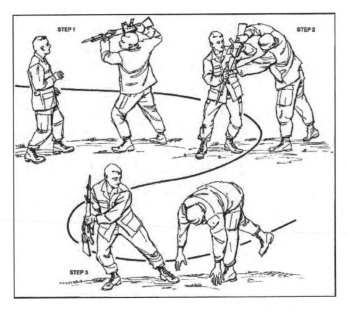

Figure 5-26: Unarmed defense against No. 6 angle of attack

The defender shifts to the outside to get off the line of attack and he grabs the weapon. Then, he pulls the attacker off balance by causing him to overextend himself (Figure 2-26, Step 2).

The defender shifts his weight backward and causes the attacker to fall, as he strips the weapon from him (Figure 5-26, Step 3).

5-7. Advanced Weapons Techniques and Training

For advanced training in weapons techniques, training partners should have the same skill level. Attackers can execute attacks along multiple angles of attack in combinations. The attacker must attack with a speed that offers the defender a challenge, but does not overwhelm him. It should not be a contest to see who can win, but a training exercise for both individuals.

a. Continued training in weapons techniques will lead to the partners' ability to engage in free-response fighting or sparring—that is, the individuals become adept enough to understand the principles of weapons attacks, defense, and movements so they can respond freely when attacking or defending from any angle.

b. Instructors must closely monitor training partners to ensure that the speed and control of the individuals does not become dangerous during advanced training practice. Proper eye protection and padding should be used, when applicable. The instructor should stress the golden rule in free-response fighting—Do unto others as you would have them do unto you.

SECTION III: OFFENSIVE TECHNIQUES

At ranges of 10 meters or more in most combat situations, small arms and grenades are the weapons of choice. However, in some scenarios, today's combat soldier must engage the enemy in confined areas, such as trench clearing or room clearing where noncombatants are present or when silence is necessary. In these instances, the bayonet or knife may be the ideal weapon to dispatch the enemy. Other than the side arm, the knife is the most lethal weapon in close-quarter combat.

5-8. Bayonet/Knife

As the bayonet is an integral part of the combat soldier's equipment, it is readily available for use as a multipurpose weapon. The bayonet produces a terrifying mental effect on the enemy when in the hands of a well-trained and confident soldier. The soldier skilled in the use of the knife also increases his ability to defend against larger opponents and multiple attackers. Both these skills increase his chances of surviving and accomplishing the mission. (Although the following paragraphs say "knife," the information also applies to bayonets.)

   a. Grips. The best way to hold the knife is either with the straight grip or the reverse grip.
      (1) Straight Grip. Grip the knife in the strong hand by forming a vee and by allowing the knife to fit naturally, as in gripping for a handshake. The handle should lay diagonally across the palm. Point the blade toward the enemy, usually with the cutting edge down. The cutting edge can also be held vertically or horizontally to the ground. Use the straight grip when thrusting and slashing.
      (2) Reverse Grip. Grip the knife with the blade held parallel with the forearm, cutting edge facing outward. This grip conceals the knife from the enemy's view. The reverse grip also affords the most power for lethal insertion. Use this grip for slashing, stabbing, and tearing.
   b. Stances. The primary stances are the knife fighter's stance and the modified stance.
      (1) Knife fighter's stance. In this stance, the fighter stands with his feet about shoulder-width apart, dominant foot toward the rear. About 70 percent of his weight is on the front foot and 30 percent on the rear foot. He stands on the balls of both feet and holds the knife with the straight grip. The other hand is held close to his body where it is ready to use, but protected (Figure 5-27).
      (2) Modified stance. The difference in the modified stance is the knife is held close to the body with the other hand held close over the knife hand to help conceal it (Figure 5-28).
   c. Range. The two primary ranges in knife fighting are long range and medium range. In long-range knife fighting, attacks consist of figure-eight slashes along the No. 1, No. 2, No. 7, and No. 8 angles of attack; horizontal slashes along the No. 3 and No. 4 angles of attack; and lunging thrusts to vital areas on the No. 5 angle of attack. Usually, the straight grip is used. In medium-range knife fighting, the reverse grip provides greater power. It is used to thrust, slash, and tear along all angles of attack.

Figure 5-27: Stance

5-9. Knife-Against-Knife Sequence

The knife fighter must learn to use all available weapons of his body and not limit himself to the knife. The free hand can be used to trap the enemy's hands to create openings in his defense. The enemy's attention will be focused on the weapon; therefore, low kicks and knee strikes will seemingly come from nowhere. The knife fighter's priority of targets are the eyes, throat, abdominal region, and extended limbs. Some knife attack sequences that can be used in training to help develop soldiers' knowledge of movements, principles, and techniques in knife fighting follow:

    a. Nos. 1 and 4 Angles. Two opponents assume the knife fighter's stance (Figure 5-29, Step 1).

        The attacker starts with a diagonal slash along the No. 1 angle of attack to the throat (Figure 5-29, Step 2).

Figure 5-28: Modified stance

He then follows through with a slash and continues with a horizontal slash back across the abdomen along the No. 4 angle of attack (Figure 5-29, Step 3).

He finishes the attack by using his entire body mass behind a lunging stab into the opponent's solar plexus (Figure 5-29, Step 4).

b. Nos. 5, 3, and 2 Angles. In this sequence, one opponent (attacker) starts an attack with a lunge along the No. 5 angle of attack. At the same time, the other opponent (defender) on the left moves his body off the line of attack, parries the attacking arm, and slices the biceps of his opponent (Figure 5-30, Step 1).

The defender slashes back across the groin along the No. 3 angle of attack (Figure 5-30, Step 2).

He finishes the attacker by continuing with an upward stroke into the armpit or throat along the No. 2 angle of attack (Figure 5-30, Step 3). Throughout this sequence, the attacker's

Figure 5-29: Nos. 1 and 4 angles

Figure 5-30: Nos. 5, 3, and 2 angles

weapon hand is controlled with the defender's left hand as he attacks with his own knife hand.

c. Low No. 5 Angle. In the next sequence, the attacker on the right lunges to the stomach along a low No. 5 angle of attack.

The defender on the left moves his body off the line of attack while parrying and slashing the wrist of the attacking knife hand as he redirects the arm (Figure 5-31, Step 1).

After he slashes the wrist of his attacker, the defender continues to move around the outside and stabs the attacker's armpit (Figure 5-31, Step 2).

He retracts his knife from the armpit, continues his movement around the attacker, and slices his hamstring (Figure 5-31, Step 3).

d. Optional Low No. 5 Angle. The attacker on the right lunges to the stomach of his opponent (the defender) along the low No. 5 angle of attack. The defender moves his body off the line of attack of the knife. Then he turns and, at the same time, delivers a slash to the attacker's throat along the No. 1 angle of attack (Figure 5-32, Step 1).

The defender immediately follows with another slash to the opposite side of the attacker's throat along the No. 2 angle of attack (Figure 5-32, Step 2).

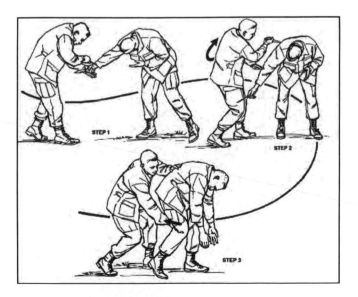

Figure 5-31: Low No. 5 angle

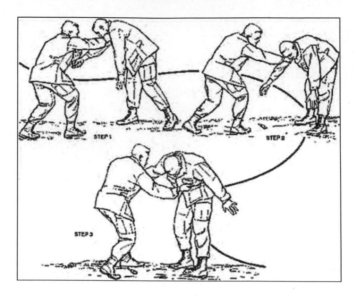

Figure 5-32: Optional low No. 5 angle

The attacker is finished as the opponent on the left (defender) continues to slice across the abdomen with a stroke along the No. 3 angle (Figure 5-32, Step 3).

5-10. Rifle with Fixed Bayonet
The principles used in fighting with the rifle and fixed bayonet are the same as when knife fighting. Use the same angles of attack and similar body movements. The principles of timing and distance remain paramount; the main difference is the extended distance provided by the length of the weapon. It is imperative that the soldier fighting with rifle and fixed bayonet use the movement of his entire body behind all of his fighting techniques—not just upper-body strength. Unit trainers should be especially conscious of stressing full body mass in motion for power and correcting all deficiencies during training. Whether the enemy is armed or unarmed, a soldier fighting with rifle and fixed bayonet must develop the mental attitude that he will survive the fight. He must continuously evaluate each moment in a fight to determine his advantages or options, as well as the enemy's. He should base his defenses on keeping his body moving and off the line of any attacks from his opponent. The soldier seeks openings in the enemy's defenses and starts his own attacks, using all available body weapons and angles of at-

**Figure 5-33: No. 1 angle of attack with rifle and
fixed bayonet**

tack. The angles of attack with rifle and fixed bayonet are shown in
Figures 5-33 through 5-39.

a. Fighting Techniques. New weapons, improved equipment,
   and new tactics are always being introduced; however, fire-
   power alone will not always drive a determined enemy from
   his position. He will often remain in defensive emplacements
   until driven out by close combat. The role of the soldier, particularly in the final phase of the assault, remains relatively
   unchanged: His mission is to close with and disable or capture
   the enemy. This mission remains the ultimate goal of all individual training. The rifle with fixed bayonet is one of the final
   means of defeating an opponent in an assault.
   (1) During infiltration missions at night or when secrecy must
       be maintained, the bayonet is an excellent silent weapon.
   (2) When close-in fighting determines the use of small-arms
       fire or grenades to be impractical, or when the situation
       does not permit the loading or reloading of the rifle, the
       bayonet is still the weapon available to the soldier.
   (3) The bayonet serves as a secondary weapon should the rifle
       develop a stoppage.

Figure 5-34: No. 2 angle of attack with rifle and fixed bayonet

Figure 5-35: No. 3 angle of attack with rifle and fixed bayonet

(4) In hand-to-hand encounters, the detached bayonet may be used as a handheld weapon.

(5) The bayonet has many nonfighting uses, such as to probe for mines, to cut vegetation, and to use for other tasks where a pointed or cutting tool is needed.

b. Development. To become a successful rifle-bayonet fighter, a soldier must be physically fit and mentally alert. A well-

**Figure 5-36: No. 4 angle of attack with rifle and
fixed bayonet**

**Figure 5-37: No. 5 angle of attack with rifle and
fixed bayonet**

rounded physical training program will increase his chances of
survival in a bayonet encounter. Mental alertness entails being
able to quickly detect and meet an opponent's attack from any
direction. Aggressiveness, accuracy, balance, and speed are essential in training as well as in combat situations. These traits
lead to confidence, coordination, strength, and endurance,
which characterize the rifle-bayonet fighter. Differences in

**Figure 5-38: High No. 5 angle of attack with rifle
and fixed bayonet**

**Figure 5-39: No. 6 angle of attack with rifle and
fixed bayonet**

individual body physique may require slight changes from
the described rifle-bayonet techniques. These variations will
be allowed if the individual's attack is effective.

c. Principles. The bayonet is an effective weapon to be used ag-
gressively; hesitation may mean sudden death. The soldier
must attack in a relentless assault until his opponent is disabled or captured. He should be alert to take advantage of

any opening. If the opponent fails to present an opening, the bayonet fighter must make one by parrying his opponent's weapon and driving his blade or rifle butt into the opponent with force.

(1) The attack should be made to a vulnerable part of the body: face, throat, chest, abdomen, or groin.

(2) In both training and combat, the rifle-bayonet fighter displays spirit by sounding off with a low and aggressive growl. This instills a feeling of confidence in his ability to close with and disable or capture the enemy.

(3) The instinctive rifle-bayonet fighting system is designed to capitalize on the natural agility and combatives movements of the soldier. It must be emphasized that precise learned movements will NOT be stressed during training.

d. Positions. The soldier holds the rifle firmly but not rigidly. He relaxes all muscles not used in a specific position; tense muscles cause fatigue and may slow him down. After proper training and thorough practice, the soldier instinctively assumes the basic positions. All positions and movements described in this manual are for right-handed men. A left-handed man, or a man who desires to learn lefthanded techniques, must use the opposite hand and foot for each phase of the movement described. All positions and movements can be executed with or without the magazine and with or without the sling attached.

(1) Attack position. This is the basic starting position (A and B, Figure 5-40) from which all attack movements originate.

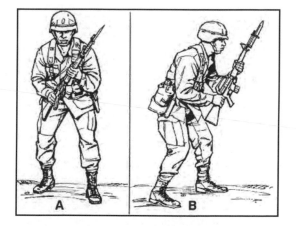

Figure 5-40: Attack position

It generally parallels a boxer's stance. The soldier assumes this position when running or hurdling obstacles. The instructor explains and demonstrates each move.

(a) Take a step forward and to the side with your left foot so that your feet are a comfortable distance apart.

(b) Hold your body erect or bend slightly forward at the waist. Flex your knees and balance your body weight on the balls of your feet. Your right forearm is roughly parallel to the ground. Hold the left arm high, generally in front of the left shoulder. Maintain eye-to-eye contact with your opponent, watching his weapon and body through peripheral vision.

(c) Hold your rifle diagonally across your body at a sufficient distance from the body to add balance and protect you from enemy blows. Grasp the weapon in your left hand just below the upper sling swivel, and place the right hand at the small of the stock. Keep the sling facing outward and the cutting edge of the bayonet toward your opponent. The command is, ATTACK POSITION, MOVE. The instructor gives the command, and the soldiers perform the movement.

(2) Relaxed position. The relaxed position (Figure 5-41) gives the soldier a chance to rest during training. It also allows him to direct his attention toward the instructor as he discusses and demonstrates the positions and movements. To assume the relaxed position from the attack position, straighten the waist and knees and lower the rifle across the front of your body by extending the arms downward. The command is, RELAX. The instructor gives the command, and the soldiers perform the movement.

e. Movements. The soldier will instinctively strike at openings and become aggressive in his attack once he has learned to relax and has developed instinctive reflexes. His movements do not have to be executed in any prescribed order. He will achieve balance in his movements, be ready to strike in any direction, and keep striking until he has disabled his opponent. There are two basic movements used throughout bayonet instruction: the whirl and the crossover. These movements develop instant reaction to commands and afford the instructor maximum control of the training formation while on the training field.

(1) Whirl movement. The whirl (Figure 5-42, Steps 1, 2, and 3), properly executed, allows the rifle bayonet fighter to meet

Figure 5-41: Relaxed position

Figure 5-42: Whirl movement

a challenge from an opponent attacking him from the rear. At the completion of a whirl, the rifle remains in the attack position. The instructor explains and demonstrates how to spin your body around by pivoting on the ball of the leading foot in the direction of the leading foot, thus facing completely about. The command is, WHIRL. The instructor gives the command, and the soldiers perform the movement.

(2) Crossover movement. While performing certain movements in rifle-bayonet training, two ranks will be moving toward each other. When the soldiers in ranks come too close to each other to safely execute additional movements, the crossover is used to separate the ranks a safe distance apart. The instructor explains and demonstrates how to move straight forward and pass your opponent so that your right shoulder passes his right shoulder, continue moving forward about six steps, halt, and without command, execute the whirl. Remain in the attack position and wait for further commands. The command is, CROSSOVER. The instructor gives the command, and the soldiers perform the movement.

NOTE: Left-handed personnel cross left shoulder to left shoulder.

(3) Attack movements. There are four attack movements designed to disable or capture the opponent: thrust, butt stroke, slash, and smash. Each of these movements may be used for the initial attack or as a follow-up should the initial movement fail to find its mark. The soldiers learn these movements separately. They will learn to execute these movements in a swift and continuous series during subsequent training. During all training, the emphasis will be on conducting natural, balanced movements to effectively damage the target. Precise, learned movements will not be stressed.

(a) Thrust. The objective is to disable or capture an opponent by thrusting the bayonet blade into a vulnerable part of his body. The thrust is especially effective in areas where movement is restricted—for example, trenches, wooded areas, or built-up areas. It is also effective when an opponent is lying on the ground or in a fighting position. The instructor explains and demonstrates how to lunge forward on your leading foot without losing your balance (Figure 5-43, Step 1) and, at the same time, drive the bayonet with great force into any unguarded part of your opponent's body.

To accomplish this, grasp the rifle firmly with both hands and pull the stock in close to the right hip; partially extend the left arm, guiding the point of the bayonet in the general direction of the opponent's body (Figure 5-43, Step 2).

Quickly complete the extension of the arms and body as the leading foot strikes the ground so that the bayonet penetrates the target (Figure 5-43, Step 3).

To withdraw the bayonet, keep your feet in place, shift your body weight to the rear, and pull rearward along the same line of penetration (Figure 5-43, Step 4).

Next, assume the attack position in preparation to continue the assault (Figure 5-43, Step 5).

This movement is taught by the numbers in three phases:

1. THRUST AND HOLD, MOVE.
2. WITHDRAW AND HOLD, MOVE.
3. ATTACK POSITION, MOVE.

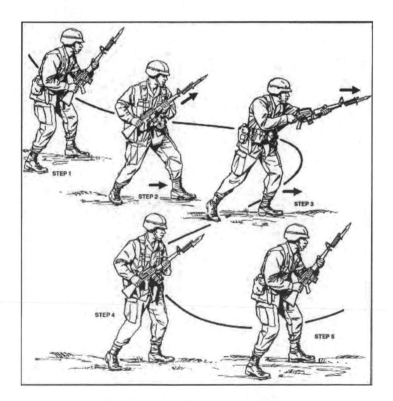

Figure 5-43: Thrust movement

At combat speed, the command is, THRUST SERIES, MOVE. Training emphasis will be placed on movement at combat speed. The instructor gives the commands, and the soldiers perform the movements.

(b) Butt stroke. The objective is to disable or capture an opponent by delivering a forceful blow to his body with the rifle butt (Figure 5-44, Steps 1, 2, 3, and 4, and Figure 5-45, Steps 1, 2, 3, and 4). The aim of the butt stroke may be the opponent's weapon or a vulnerable portion of his body. The butt stroke may be vertical, horizontal, or somewhere between the two planes. The instructor explains and demonstrates how to step forward with your trailing foot and, at the same time using your left hand as a pivot, swing the rifle in an arc and drive the rifle butt into your opponent. To recover, bring your trailing foot forward and assume the attack position. The movement is taught by the numbers in two phases:

1. BUTT STROKE TO THE (head, groin, kidney) AND HOLD, MOVE.
2. ATTACK POSITION, MOVE.

At combat speed, the command is, BUTT STROKE TO THE (head, groin, kidney) SERIES, MOVE. Training emphasis will be placed on movement at combat speed. The instructor gives the commands, and the soldiers perform the movement.

(c) Slash. The objective is to disable or capture the opponent by cutting him with the blade of the bayonet. The instructor

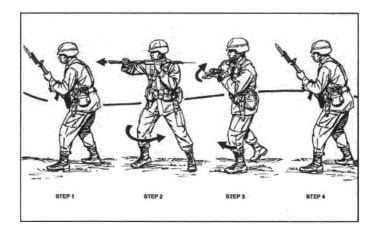

Figure 5-44: Butt stroke to the head

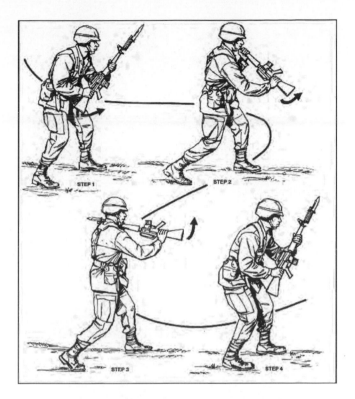

Figure 5-45: Butt stroke to the groin

explains and demonstrates how to step forward with your lead foot (Figure 5-46, Step 1).

At the same time, extend your left arm and swing the knife edge of your bayonet forward and down in a slashing arc (Figure 5-46, Steps 2 and 3).

To recover, bring your trailing foot forward and assume the attack position (Figure 5-46, Step 4).

This movement is taught by the number in two phases:

1. SLASH AND HOLD, MOVE.
2. ATTACK POSITION, MOVE.

At combat speed, the command is, SLASH SERIES, MOVE. Training emphasis will be placed on movement at combat speed. The instructor gives the commands, and the soldiers perform the movements.

(d) Smash. The objective is to disable or capture an opponent by smashing the rifle butt into a vulnerable part of his body.

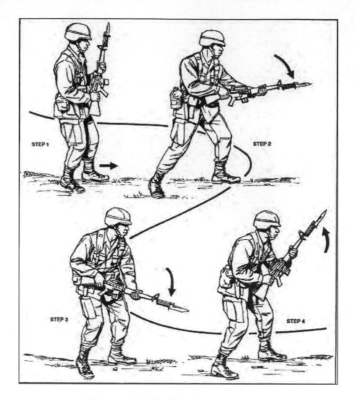

Figure 5-46: Slash movement

The smash is often used as a follow-up to a butt stroke and is also effective in wooded areas and trenches when movement is restricted. The instructor explains and demonstrates how to push the butt of the rifle upward until horizontal (Figure 5-47, Step 1) and above the left shoulder with the bayonet pointing to the rear, sling up (Figure 5-47, Step 2). The weapon is almost horizontal to the ground at this time.

Step forward with the trailing foot, as in the butt stroke, and forcefully extend both arms, slamming the rifle butt into the opponent (Figure 5-47, Step 3).

To recover, bring your trailing foot forward (Figure 5-47, Step 4) and assume the attack position (Figure 5-47, Step 5). This movement is taught by the numbers in two phases:

1. SMASH AND HOLD, MOVE.
2. ATTACK POSITION, MOVE.

At combat speed, the command is, SMASH SERIES, MOVE. Training emphasis will be placed on movement at

Figure 5-47: Smash movement

combat speed. The instructor gives the commands, and the soldiers perform the movements.

(4) Defensive movements. At times, the soldier may lose the initiative and be forced to defend himself. He may also meet an opponent who does not present a vulnerable area to attack. Therefore, he must make an opening by initiating a parry or block movement, then follow up with a vicious attack. The follow-up attack is immediate and violent.

⚠ CAUTION

TO MINIMIZE WEAPON DAMAGE WHILE USING BLOCKS AND PARRIES, LIMIT WEAPON-TO-WEAPON CONTACT TO HALF SPEED DURING TRAINING.

(a) Parry movement. The objective is to counter a thrust, throw the opponent off balance, and hit a vulnerable area of his body. Timing, speed, and judgment are essential factors in these movements. The instructor explains and demonstrates how to—

- Parry right. If your opponent carries his weapon on his left hip (left-handed), you will parry it to your right. In execution, step forward with your leading foot (Figure 5-48, Step 1), strike the opponent's rifle (Figure 5-48, Step 2), deflecting it to your right (Figure 5-48, Step 3), and follow up with a thrust, slash, or butt stroke.
- Parry left. If your opponent carries his weapon on his right hip (right-handed), you will parry it to your left. In execution, step forward with your leading foot (Figure 5-49, Step 1), strike the opponent's rifle (Figure 5-49, Step 2), deflecting it to your left (Figure 5-49, Step 3), and follow up with a thrust, slash, or butt stroke.

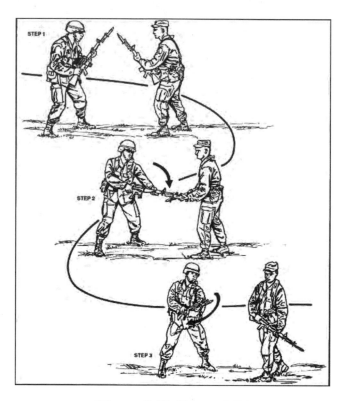

Figure 5-48: Parry right

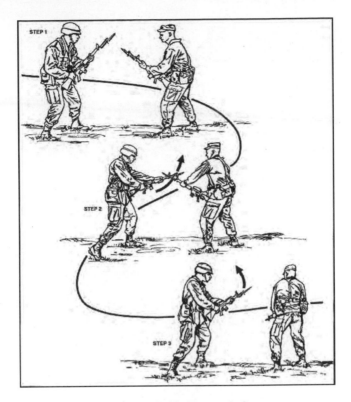

Figure 5-49: Parry left

A supplementary parry left is the follow-up attack (Figure 5-50, Steps 1, 2, 3, 4, and 5).
- Recovery. Immediately return to the attack position after completing each parry and follow-up attack.

The movement is taught by the numbers in three phases:
1. PARRY RIGHT (OR LEFT), MOVE.
2. THRUST MOVE.
3. ATTACK POSITION, MOVE.

At combat speed, the command is, PARRY RIGHT (LEFT) or PARRY (RIGHT OR LEFT) WITH FOLLOW-UP ATTACK. The instructor gives the commands, and the soldiers perform the movements.

(b) Block. When surprised by an opponent, the block is used to cut off the path of his attack by making weapon-to-weapon contact. A block must always be followed immediately with a vicious attack. The instructor explains and demonstrates

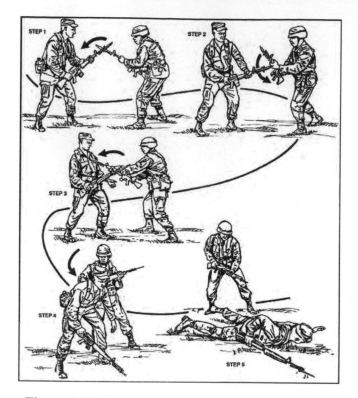

**Figure 5-50: Parry left, slash, with follow-up butt
stroke to kidney region**

how to extend your arms using the center part of your rifle
as the strike area, and cut off the opponent's attack by making weapon-to-weapon contact. Strike the opponent's weapon
with enough power to throw him off balance.

- High block (Figure 5-51, Steps 1, 2, and 3). Extend your
  arms upward and forward at a 45- degree angle. This action
  deflects an opponent's slash movement by causing his bayonet or upper part of his rifle to strike against the center
  part of your rifle.
- Low block (Figure 5-52, Steps 1, 2, and 3). Extend your
  arms downward and forward about 15 degrees from your
  body. This action deflects an opponent's butt stroke aimed
  at the groin by causing the lower part of his rifle stock to
  strike against the center part of your rifle.
- Side block (Figure 5-53, Steps 1 and 2). Extend your arms
  with the left hand high and right hand low, thus holding

Figure 5-51: High block against slash

**Figure 5-52: Low block against butt
stroke to groin**

Figure 5-53: Side block against butt stroke

the rifle vertical. This block is designed to stop a butt stroke aimed at your upper body or head. Push the rifle to your left to cause the butt of the opponent's rifle to strike the center portion of your rifle.

- Recovery. Counterattack each block with a thrust, butt stroke, smash, or slash.

Blocks are taught by the numbers in two phases:

1. HIGH (LOW) or (SIDE) BLOCK.
2. ATTACK POSITION, MOVE.

At combat speed, the command is the same. The instructor gives the commands, and the soldiers perform the movement.

(5) Modified movements. Two attack movements have been modified to allow the rifle-bayonet fighter to slash or thrust an opponent without removing his hand from the pistol grip of the M16 rifle should the situation dictate.

    (a) The modified thrust (Figure 5-54, Steps 1 and 2) is identical to the thrust (as described in paragraph (3) (a)) with the exception of the right hand grasping the pistol grip.

    (b) The modified slash (Figure 5-55, Steps 1, 2, 3, and 4) is identical to the slash (as described in paragraph (3 (c)) with the exception of the right hand grasping the pistol grip.

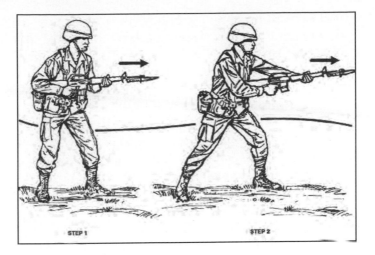

Figure 5-54: Modified thrust

Figure 5-55: Modified slash

(6) Follow-up movements. Follow-up movements are attack movements that naturally follow from the completed position of the previous movement. If the initial thrust, butt stroke, smash, or slash fails to make contact with the opponent's body, the soldier should instinctively follow up with additional movements until he has disabled or captured the opponent. It is important to follow up the initial attack with another aggressive action so the initiative is not lost. The instructor explains and demonstrates how instinct should govern your selection of a specific follow-up movement. For example—

- PARRY LEFT, BUTT STROKE TO THE HEAD, SMASH, SLASH, ATTACK POSITION.
- PARRY LEFT, SLASH, BUTT STROKE TO THE KIDNEY, ATTACK POSITION.
- PARRY RIGHT THRUST, BUTT STROKE TO THE GROIN, SLASH, ATTACK POSITION.

Two examples of commands using follow-up movements are—

- PARRY LEFT (soldier executes), THRUST (soldier executes), BUTT STROKE TO THE HEAD (soldier executes), SMASH (soldier executes), SLASH (soldier executes), ATTACK POSITION (soldier assumes the attack position).
- THRUST (soldier executes), THRUST (soldier executes), THRUST (soldier executes), BUTT STROKE TO THE GROIN (soldier executes), SLASH (soldier executes), ATTACK POSITION (soldier assumes the attack position).

All training will stress damage to the target and violent action, using natural movements as opposed to precise, stereotyped movements. Instinctive, aggressive action and balance are the keys to offense with the rifle and bayonet.

NOTE: For training purposes, the instructor may and should mix up the series of movements.

SECTION IV: FIELD-EXPEDIENT WEAPONS

To survive, the soldier in combat must be able to deal with any situation that develops. His ability to adapt any nearby object for use as a weapon in a win-or-die situation is limited only by his ingenuity and resourcefulness. Possible weapons, although not discussed herein, include ink pens or pencils; canteens tied to string to be swung; snap links at the end of sections of rope; kevlar helmets; sand, rocks, or liquids thrown into the enemy's eyes; or radio antennas. The following techniques demonstrate a few expedient weapons that are readily available to most soldiers for defense and counterattack against the bayonet and rifle with fixed bayonet.

5-11. Entrenching Tool

Almost all soldiers carry the entrenching tool. It is a versatile and formidable weapon when used by a soldier with some training. It can be used in its straight position—locked out and fully extended—or with its blade bent in a 90-degree configuration.

    a. To use the entrenching tool against a rifle with fixed bayonet, the attacker lunges with a thrust to the stomach of the defender along a low No. 5 angle of attack (Figure 5-56, Step 1).

        The defender moves just outside to avoid the lunge and meets the attacker's arm with the blade of the fully extended entrenching tool (Figure 5-56, Step 2).

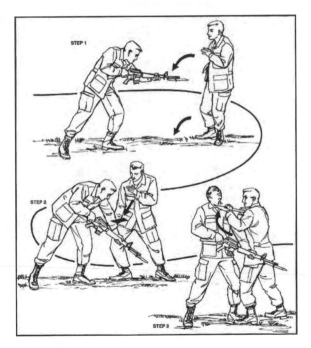

Figure 5-56: Entrenching tool against rifle with fixed bayonet

The defender gashes all the way up the attacker's arm with the force of both body masses coming together. The hand gripping the entrenching tool is given natural protection from the shape of the handle. The defender continues pushing the blade of the entrenching tool up and into the throat of the attacker, driving him backward and downward (Figure 5-56, Step 3).

b. An optional use of entrenching tool against a rifle with fixed bayonet is for the attacker to lunge to the stomach of the defender (Figure 5-57, Step 1).

The defender steps to the outside of the line of attack at 45 degrees to avoid the weapon. He then turns his body and strikes downward onto the attacking arm (on the radial nerve) with the blade of the entrenching tool (Figure 5-57, Step 2).

He drops his full body weight down with the strike, and the force causes the attacker to collapse forward. The defender then strikes the point of the entrenching tool into the jugular notch, driving it deeply into the attacker (Figure 5-57, Step 3).

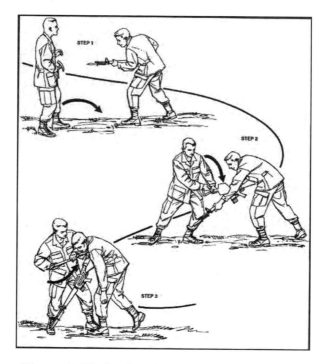

Figure 5-57: Optional use of entrenching tool against rifle with fixed bayonet

c. In the next two sequences, the entrenching tool is used in the bent configuration—that is, the blade is bent 90 degrees to the handle and locked into place.

(1) The attacker tries to stick the bayonet into the chest of the defender (Figure 5-58, Step 1).

When the attack comes, the defender moves his body off the line of attack by stepping to the outside. He allows his weight to shift forward and uses the blade of the entrenching tool to drag along the length of the weapon, scraping the attacker's arm and hand (Figure 5-58, Step 2). The defender's hand is protected by the handle's natural design.

He continues to move forward into the attacker, strikes the point of the blade into the jugular notch, and drives it downward (Figure 5-58, Step 3).

(2) The attacker lunges with a fixed bayonet along the No. 5 angle of attack (Figure 5-59, Step 1).

The defender then steps to the outside to move off the line of attack and turns; he strikes the point of the blade of

Figure 5-58: Entrenching tool in bent configuration

**Figure 5-59: Optional use of entrenching
tool in bent configuration**

the entrenching tool into the side of the attacker's throat
(Figure 5-59, Step 2).

5-12. Three-Foot Stick

Since a stick can be found almost anywhere, a soldier should know
its uses as a field-expedient weapon. The stick is a versatile weapon;
its capability ranges from simple prisoner control to lethal combat.

a. Use a stick about 3 feet long and grip it by placing it in the
   vee formed between the thumb and index finger, as in a
   handshake. It may also be grasped by two hands and used in
   an unlimited number of techniques. The stick is not held at the
   end, but at a comfortable distance from the butt end.

b. When striking with the stick, achieve maximum power by
   using the entire body weight behind each blow. The desired
   point of contact of the weapon is the last 2 inches at the tip of
   the stick. The primary targets for striking with the stick are
   the vital body points in Chapter 4. Effective striking points are
   usually the wrist, hand, knees, and other bony protuberances.

Soft targets include the side of the neck, jugular notch, solar plexus, and various nerve motor points. Attack soft targets by striking or thrusting the tip of the stick into the area. Three basic methods of striking are—

(1) Thrusting. Grip the stick with both hands and thrust straight into a target with the full body mass behind it.

(2) Whipping. Hold the stick in one hand and whip it in a circular motion; use the whole body mass in motion to generate power.

(3) Snapping. Snap the stick in short, shocking blows, again with the body mass behind each strike.

c. When the attacker thrusts with a knife to the stomach of the defender with a low No. 5 angle of attack, the defender moves off the line of attack to the outside and strikes vigorously downward onto the attacking wrist, hand, or arm (Figure 5-60, Step 1).

The defender then moves forward, thrusts the tip of the stick into the jugular notch of the attacker (Figure 5-60, Step 2),

Figure 5-60: Three-foot stick against knife

and drives him to the ground with his body weight—not his upper body strength (Figure 5-60, Step 3).

d. When using a three-foot stick against a rifle with fixed bayonet, the defender grasps the stick with two hands, one at each end, as the attacker thrusts forward to the chest (Figure 5-61, Step 1).

He steps off the line of attack to the outside and redirects the weapon with the stick (Figure 5-61, Step 2).

He then strikes forward with the forearm into the attacker's throat (Figure 5-61, Step 3). The force of the two body weights coming together is devastating. The attacker's neck is trapped in the notch formed by the stick and the defender's forearm.

Using the free end of the stick as a lever, the defender steps back and uses his body weight to drive the attacker to the ground. The leverage provided by the stick against the neck creates a tremendous choke with the forearm, and the attacker loses control completely (Figure 5-61, Step 4).

**Figure 5-61: Three-foot stick against rifle
with fixed bayonet**

5-13. Three-Foot Rope

A section of rope about 3 feet long can provide a useful means of self-defense for the unarmed combat soldier in a hand-to-hand fight. Examples of field-expedient ropes are a web belt, boot laces, a portion of a 120-foot nylon rope or sling rope, or a cravat rolled up to form a rope. Hold the rope at the ends so the middle section is rigid enough to almost serve as a stick-like weapon, or the rope can be held with the middle section relaxed, and then snapped by vigorously pulling the hands apart to strike parts of the enemy's body, such as the head or elbow joint, to cause serious damage. It can also be used to entangle limbs or weapons held by the opponent, or to strangle him.

a. When the attacker lunges with a knife to the stomach (Figure 5-62, Step 1), the defender moves off the line of attack 45 degrees to the outside.

He snaps the rope downward onto the attacking wrist, redirecting the knife (Figure 5-62, Step 2).

Then, he steps forward, allowing the rope to encircle the attacker's neck (Figure 5-62, Step 3).

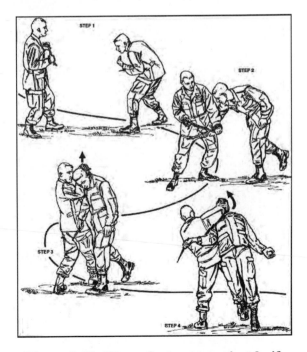

Figure 5-62: Three-foot rope against knife

He continues to turn his body and sinks his weight to drop the attacker over his hip (Figure 5-62, Step 4).

b. When the attacker thrusts with a fixed bayonet (Figure 5-63, Step 1), the defender moves off the line of attack and uses the rope to redirect the weapon (Figure 5-63, Step 2).

Then, he moves forward and encircles the attacker's throat with the rope (Figure 5-63, Step 3). He continues moving to unbalance the attacker and strangles him with the rope (Figure 5-63, Step 4).

c. The 3-foot rope can also be a useful tool against an unarmed opponent. The defender on the left prepares for an attack by gripping the rope between his hands (Figure 5-64, Step 1).

When the opponent on the right attacks, the defender steps completely off the line of attack and raises the rope to strike the attacker's face (Figure 5-64, Step 2).

He then snaps the rope to strike the attacker either across the forehead, just under the nose, or under the chin by jerking his hands forcefully apart. The incoming momentum of the

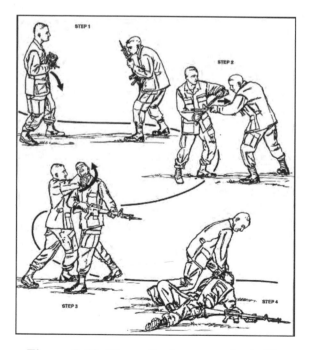

**Figure 5-63: Three-foot rope against rifle
with fixed bayonet**

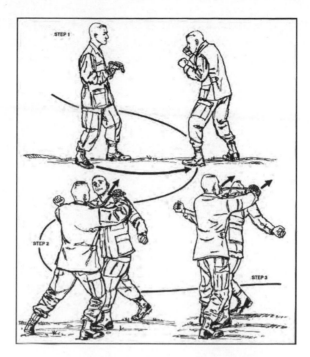

**Figure 5-64: Three-foot rope against
unarmed opponent**

attacker against the rope will snap his head backward, will probably break his neck, or will at least knock him off his feet (Figure 5-64, Step 3).

5-14. Six-Foot Pole

Another field-expedient weapon that can mean the difference between life and death for a soldier in an unarmed conflict is a pole about 6 feet long. Examples of poles suitable for use are mop handles, pry bars, track tools, tent poles, and small trees or limbs cut to form a pole. A soldier skilled in the use of a pole as a weapon is a formidable opponent. The size and weight of the pole requires him to move his whole body to use it effectively. Its length gives the soldier an advantage of distance in most unarmed situations. There are two methods usually used in striking with a pole:

    a. Swinging. Becoming effective in swinging the pole requires skilled body movement and practice. The greatest power is developed by striking with the last 2 inches of the pole.

b. Thrusting. The pole is thrust straight along its axis with the user's body mass firmly behind it.

   (1) An attacker tries to thrust forward with a fixed bayonet (Figure 5-65, Step 1).

      The defender moves his body off the line of attack; he holds the tip of the pole so that the attacker runs into it from his own momentum. He then aims for the jugular notch and anchors his body firmly in place so that the full force of the attack is felt at the attacker's throat (Figure 5-65, Step 2).

   (2) The defender then shifts his entire body weight forward over his lead foot and drives the attacker off his feet (Figure 5-65, Step 3).

NOTE: During high stress, small targets, such as the throat, may be difficult to hit. Good, large targets include the solar plexus and hip/thigh joint.

Figure 5-65: Thrusting with 6-foot pole

CHAPTER 6

SENTRY REMOVAL

Careful planning, rehearsal, and execution are vital to the success of a mission that requires the removal of a sentry. This task may be necessary to gain access to an enemy location or to escape confinement.

6-1. Planning Considerations

A detailed schematic of the layout of the area guarded by sentries must be available. Mark known and suspected locations of all sentries. It will be necessary—

   a. To learn the schedule for the changing of the guards and the checking of the posts.
   b. To learn the guard's meal times. It may be best to attack a sentry soon after he has eaten when his guard is lowered. Another good time to attack the sentry is when he is going to the latrine.
   c. To post continuous security.
   d. To develop a contingency plan.
   e. To plan infiltration and exfiltration routes.
   f. To carefully select personnel to accomplish the task.
   g. To carry the least equipment necessary to accomplish the mission because silence, stealth, and ease of movement are essential.
   h. To conceal or dispose of killed sentries.

6-2. Rehearsals

Reproduce and rehearse the scenario of the mission as closely as possible to the execution phase.

Conduct the rehearsal on similar terrain, using sentries, the time schedule, and the contingency plan. Use all possible infiltration and exfiltration routes to determine which may be the best.

141

6-3. Execution

When removing a sentry, the soldier uses his stalking skills to approach the enemy undetected. He must use all available concealment and keep his silhouette as low as possible.

a. When stepping, the soldier places the ball of his lead foot down first and checks for stability and silence of the surface to be crossed. He then lightly touches the heel of his lead foot. Next, he transfers his body weight to his lead foot by shifting his body forward in a relaxed manner. With the weight on the lead foot, he can bring his rear foot forward in a similar manner.

b. When approaching the sentry, the soldier synchronizes his steps and movement with the enemy's, masking any sounds. He also uses background noises to mask his sounds. He can even follow the sentry through locked doors this way. He is always ready to strike immediately if he is discovered. He focuses his attention on the sentry's head since that is where the sentry generates all of his movement and attention. However, it is important not to stare at the enemy because he may sense the stalker's presence through a sixth sense. He focuses on the sentry's movements with his peripheral vision. He gets to within 3 or 4 feet and at the proper moment makes the kill as quickly and silently as possible.

c. The attacker's primary focus is to summon all of his mental and physical power to suddenly explode onto the target. He maintains an attitude of complete confidence throughout the execution. He must control fear and hesitation because one instant of hesitation could cause his defeat and compromise the entire mission.

6-4. Psychological Aspects

Killing a sentry is completely different than killing an enemy soldier while engaged in a firefight. It is a cold and calculated attack on a specific target. After observing a sentry for hours, watching him eat or look at his wife's photo, an attachment is made between the stalker and the sentry. Nonetheless, the stalker must accomplish his task efficiently and brutally. At such close quarters, the soldier literally feels the sentry fight for his life. The sights, sounds, and smells of this act are imprinted in the soldier's mind; it is an intensely personal experience. A soldier who has removed a sentry should be observed for signs of unusual behavior for four to seven days after the act.

6-5. Techniques

The following techniques are proven and effective ways to remove sentries. A soldier with moderate training can execute the proper technique for his situation, when he needs to.

a. Brachial Stun, Throat Cut. This technique relies on complete mental stunning to enable the soldier to cut the sentry's throat, severing the trachea and carotid arteries. Death results within 5 to 20 seconds. Some sounds are emitted from the exposed trachea, but the throat can be cut before the sentry can recover from the effect of the stunning strike and cry out. The soldier silently approaches to within striking range of the sentry (Figure 6-1, Step 1). The soldier strikes the side of the sentry's neck with the knife butt or a hammer fist strike (Figure 6-1, Step 2), which completely stuns the sentry for three to seven seconds. He then uses his body weight to direct the sentry's body to sink in one direction and uses his other hand to twist the sentry's head to the side, deeply cutting the throat across the front in the opposite direction (Figure 6-1, Step 3).

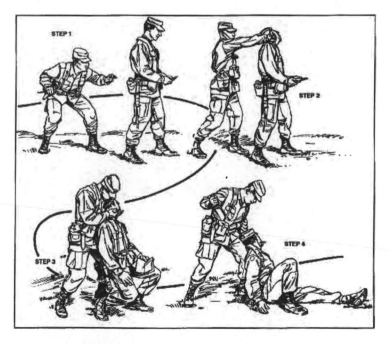

Figure 6-1: Brachial stun, throat cut

He executes the entire length of the blade in a slicing motion. The sentry's sinking body provides most of the force—not the soldier's upper-arm strength (Figure 6-1, Step 4).

b. Kidney Stab, Throat Cut. This technique relies on a stab to the kidney (Figure 6-2, Step 1) to induce immediate shock. The kidney is relatively accessible and by inducing shock with such a stab, the soldier has the time to cut the sentry's throat. The soldier completes his stalk and stabs the kidney by pulling the sentry's balance backward and downward and inserts the knife upward against his weight. The sentry will possibly gasp at this point, but shock immediately follows. By using the sentry's body weight that is falling downward and turning, the soldier executes a cut across the front of the throat (Figure 6-2, Step 2). This completely severs the trachea and carotid arteries.

Figure 6-2: Kidney stab, throat cut

c. Pectoral Muscle Strike, Throat Cut. The stun in this technique is produced by a vigorous strike to the stellate ganglia nerve center at the top of the pectoral muscle (Figure 6-3, Step 1).

The strike is delivered downward with the attacker's body weight. Use the handle of the knife for impact. Care should be taken to avoid any equipment worn by the sentry that could obstruct the strike. Do not try this technique if the sentry is wearing a ballistic vest or bulky LCE. The sentry is unable to make a sound or move if the stun is properly delivered. The throat is then cut with a vertical stab downward into the sub-clavian artery at the junction of the neck and clavicle (Figure 6-3, Step 2). Death comes within 3 to 10 seconds, and the sentry is lowered to the ground.

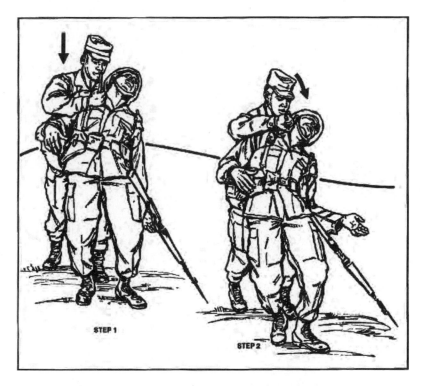

Figure 6-3: Pectoral muscle strike, throat cut

d. Nose Pinch, Mouth Grab, Throat Cut. In this technique, completely pinch off the sentry's mouth and nose to prevent any outcry. Then cut his throat or stab his subclavian artery (Figure 6-4). The danger with this technique is that the sentry can resist until he is killed, although he cannot make a sound.

Figure 6-4: Nose pinch, mouth grab, throat cut

e. Crush Larynx, Subclavian Artery Stab. Crush the sentry's larynx by inserting the thumb and two or three fingers behind his larynx, then twisting and crushing it. The subclavian artery can be stabbed at the same time with the other hand (Figure 6-5).

f. Belgian Takedown. In the Belgian takedown technique, the unsuspecting sentry is knocked to the ground and kicked in the groin, inducing shock. The soldier can then kill the sentry by any proper means. Since surprise is the essential element of this technique, the soldier must use effective stalking techniques (Figure 6-6, Step 1). To initiate his attack, he grabs both

Figure 6-5: Crush larynx, subclavian artery stab

of the sentry's ankles (Figure 6-6, Step 2). Then he heaves his body weight into the hips of the sentry while pulling up on the ankles. This technique slams the sentry to the ground on his face. Then, the soldier follows with a kick to the groin (Figure 6-6, Step 3).

g. Neck Break With Sentry Helmet. The soldier can break the sentry's neck by vigorously snatching back and down on the sentry's helmet (Figure 6-7, Step 1) while forcing the sentry's body weight forward with a knee strike (Figure 6-7, Step 2). The chin strap of the helmet must be fastened for this technique to work.

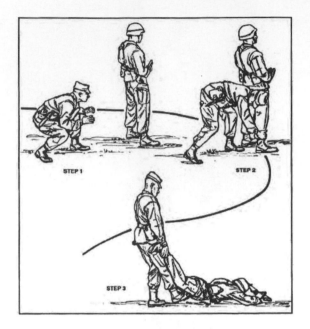

Figure 6-6: Belgian takedown

Figure 6-7: Break neck with sentry helmet

h. Knockout With Helmet. The sentry's helmet is stripped from his head and used by the soldier to knock him out (Figure 6-8, Step 1). The soldier uses his free hand to stabilize the sentry during the attack. This technique can only be used when the sentry's chin strap is loose. The preferred target area for striking with the helmet is at the base of the skull or on the temple (Figure 6-8, Step 2).

Figure 6-8: Knockdown with helmet

i. The Garrote. In this technique, use a length of wire, cord, rope, or webbed belt to takeout a sentry. Silence is not guaranteed, but the technique is effective if the soldier is unarmed and must escape from a guarded area. The soldier carefully stalks the sentry from behind with his garrote ready (Figure 6-9, Step 1). He loops the garrote over the sentry's head across the throat (Figure 6-9, Step 2) and forcefully pulls him backward as he turns his own body to place his hips in low against the hips of the sentry. The sentry's balance is already taken at this point, and the garrote becomes crossed around the sentry's throat when the turn is made. The sentry is thrown over the soldier's shoulder and killed by strangling or breaking his neck (Figure 6-9, Step 3).

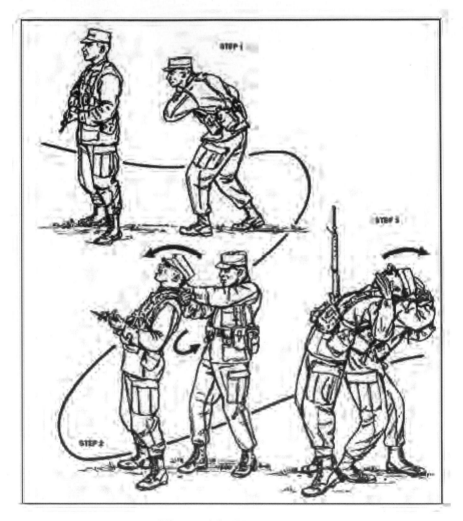

Figure 6-9: The garrote

CHAPTER 7

COVER, CONCEALMENT, AND CAMOUFLAGE

GENERAL

In a survival situation where you are in hostile territory, if the enemy can see you, he can hit you with his fire. So you must be concealed from enemy observation and have cover from enemy fire.

When the terrain does not provide natural cover and concealment, you must prepare your cover and use natural and man-made materials to camouflage yourself, your equipment, and your position. This chapter provides guidance on the preparation and use of cover, concealment, and camouflage.

COVER

Cover gives protection from bullets, fragments of exploding rounds, flame, nuclear effects, and biological and chemical agents. Cover can also conceal you from enemy observation. Cover can be natural or man-made

Natural cover includes such things as logs, trees, stumps, ravines, and hollows. Man-made cover includes such things as fighting positions, trenches, walls, rubble, and craters. Even the smallest depression or fold in the ground can give some cover. Look for and use every bit of cover the terrain offers.

In combat, you need protection from enemy direct and indirect fire.

To get this protection in the defense, build a fighting position (man-made cover) to add to the natural cover afforded by the terrain.

To get protection from enemy fire in the offense or when moving, use routes that put cover between you and the places where the enemy is known or thought to be. Use ravines, gullies, hills, wooded areas, walls, and other cover to keep the enemy from seeing and firing at you. Avoid open areas, and do not skyline yourself on hilltops and ridges.

TYPES OF COVER

FIGHTING POSITION WITH COVER

TROOPS MOVING ALONG A RAVINE

CONCEALMENT

Concealment is anything that hides you from enemy observation. Concealment does not protect you from enemy fire. Do not think that you are protected from the enemy's fire just because you are concealed. Concealment, like cover, can also be natural or man-made.

Natural concealment includes such things as bushes, grass, trees, and shadows. If possible, natural concealment should not be disturbed. Man-made concealment includes such things as battle-dress uniforms, camouflage nets, face paint, and natural material that has been moved from its original location. Man-made concealment must blend into the natural concealment provided by the terrain.

Light discipline, noise discipline, movement discipline, and the use of camouflage contribute to concealment. Light discipline is controlling the use of lights at night by such things as not smoking in the open, not walking around with a flashlight on, and not using vehicle headlights. Noise discipline is taking action to deflect sounds generated by your unit (such as operating equipment) away from the enemy and, when possible, using methods to communicate that do not generate sounds (arm-and-hand signals). Movement discipline is

such things as not moving about fighting positions unless necessary, and not moving on routes that lack cover and concealment. In the defense, build a well-camouflaged fighting position and avoid moving about. In the offense, conceal yourself and your equipment with camouflage and move in woods or on terrain that gives concealment. Darkness cannot hide you from enemy observation in either offense or defense. The enemy's night vision devices and other detection means let them find you in both daylight and darkness.

CAMOUFLAGE

Camouflage is anything you use to keep yourself, your equipment, and your position from looking like what they are. Both natural and man-made material can be used for camouflage.

Change and improve your camouflage often. The time between changes and improvements depends on the weather and on the material used. Natural camouflage will often die, fade, or otherwise lose its effectiveness. Likewise, man-made camouflage may wear off or fade. When those things happen, you and your equipment or position may not blend with the surroundings. That may make it easy for the enemy to spot you.

CAMOUFLAGE CONSIDERATIONS

Movement draws attention. When you give arm-and-hand signals or walk about your position, your movement can be seen by the naked eye at long ranges. In the defense, stay low and move only when necessary. In the offense, move only on covered and concealed routes.

POSITION IN COVER AND CONCEALMENT ON A HILLSIDE

FRONTAL COVER

FIRING PORTS

Positions must not be where the enemy expects to find them. Build positions on the side of a hill, away from road junctions or lone buildings, and in covered and concealed places. Avoid open areas.

Outlines and shadows may reveal your position or equipment to air or ground observers. Outlines and shadows can be broken up with camouflage. When moving, stay in the shadows when possible.

Shine may also attract the enemy's attention. In the dark, it may be a light such as a burning cigarette or flashlight. In daylight, it can be reflected light from polished surfaces such as shiny mess gear, a worn helmet, a windshield, a watch crystal and band, or exposed skin. A light, or its reflection, from a position may help the enemy detect the position. To reduce shine, cover your skin with clothing and face paint. However, in a nuclear attack, darkly painted skin can absorb more thermal energy and may burn more readily than bare skin. Also, dull the surfaces of equipment and vehicles with paint, mud, or some type of camouflage material.

Shape is outline or form. The shape of a helmet is easily recognized. A human body is also easily recognized. Use camouflage and concealment to breakup shapes and blend them with their surroundings. Be careful not to overdo it.

HELMET CAMOUFLAGED

The colors of your skin, uniform, and equipment may help the enemy detect you if the colors contrast with the background. For example, a green uniform will contrast with snow-covered terrain. Camouflage yourself and your equipment to blend with the surroundings.

SOLDIER IN ARCTIC CAMOUFLAGE

Dispersion is the spreading of men, vehicles, and equipment over a wide area. It is usually easier for the enemy to detect soldiers when they are bunched. So, spread out. The distance between you and your fellow soldier will vary with the terrain, degree of visibility, and enemy situation. Distances will normally be set by unit leaders or by a unit's standing operating procedure (SOP).

HOW TO CAMOUFLAGE

Before camouflaging, study the terrain and vegetation of the area in which you are operating. Then pick and use the camouflage material that best blends with that area. When moving from one area to another, change camouflage as needed to blend with the surroundings. Take grass, leaves, brush, and other material from your location and apply it to your uniform and equipment and put face paint on your skin.

Fighting Positions. When building a fighting position, camouflage it and the dirt taken from it. Camouflage the dirt used as frontal, flank, rear, and overhead cover. Also camouflage the bottom of the hole to prevent detection from the air. If necessary, take excess dirt away from the position (to the rear).

Do not over camouflage. Too much camouflage material may actually disclose a position. Get your camouflage material from a wide area. An area stripped of all or most of its vegetation may draw

CAMOUFLAGED SOLDIERS

attention. Do not wait until the position is complete to camouflage it. Camouflage the position as you build.

CAMOUFLAGED POSITION BEING IMPROVED

Do not leave shiny or light-colored objects lying about. Hide mess kits, mirrors, food containers, and white underwear and towels. Do not remove your shirt in the open. Your skin may shine and be seen. Never use fires where there is a chance that the flame will be seen or the smoke will be smelled by the enemy. Also, cover up tracks and other signs of movement.

When camouflage is complete, inspect the position from the enemy's side. This should be done from about 35 meters forward of the position. Then check the camouflage periodically to see that it stays natural-looking and conceals the position. When the camouflage becomes ineffective, change and improve it.

Helmets. Camouflage your helmet with the issue helmet cover or make a cover of cloth or burlap that is colored to blend with the terrain. The cover should fit loosely with the flaps folded under the helmet or left hanging. The hanging flaps may break up the helmet outline. Leaves, grass, or sticks can also be attached to the cover. Use camouflage bands, strings, burlap strips, or rubber bands to hold

those in place. If there is no material for a helmet cover, disguise and dull helmet surface with irregular patterns of paint or mud.

Uniforms. Most uniforms come already camouflaged. However, it may be necessary to add more camouflage to make the uniform blend better with the surroundings. To do this, put mud on the uniform or attach leaves, grass, or small branches to it. Too much camouflage, however, may draw attention.

When operating on snow-covered ground, overwhites (if issued) to help blend with the snow. If overwhites are not issued, use white cloth, such as white bedsheets, to get the same effect.

Skin. Exposed skin reflects light and may wear draw the enemy's attention. Even very dark skin, because of its natural oil, will reflect light. Use the following methods when applying camouflage face paint to camouflage the skin.

	SKIN COLOR	SHINE AREAS	SHADOW AREAS
CAMOUFLAGE MATERIAL	LIGHT OR DARK	FOREHEAD, CHEEKBONES, EARS, NOSE AND CHIN	AROUND EYES, UNDER NOSE, AND UNDER CHIN
LOAM AND LIGHT GREEN STICK	ALL TROOPS USE IN AREAS WITH GREEN VEGETATION	USE LOAM	USE LIGHT GREEN
SAND AND LIGHT GREEN STICK	ALL TROOPS USE IN AREAS LACKING GREEN VEGETATION	USE LIGHT GREEN	USE SAND
LOAM AND WHITE	ALL TROOPS USE ONLY IN SNOW-COVERED TERRAIN	USE LOAM	USE WHITE
BURNT CORK, BARK CHARCOAL, OR LAMP BLACK	ALL TROOPS, IF CAMOUFLAGE STICKS NOT AVAILABLE	USE	DO NOT USE
LIGHT-COLOR MUD	ALL TROOPS, IF CAMOUFLAGE STICKS NOT AVAILABLE	DO NOT USE	USE

When applying camouflage stick to your skin, work with a buddy (in pairs) and help each other. Apply a two-color combination of camouflage stick in an irregular pattern. Paint shiny areas (forehead, cheekbones, nose, ears, and chin) with a dark color. Paint shadow areas (around the eyes, under the nose, and under the chin) with a light color. In addition to the face, paint the exposed skin on the back of the neck, arms, and hands. Palms of hands are not normally camouflaged if arm-and-hand signals are to be used. Remove all jewelry to further reduce shine or reflection.

When camouflage sticks are not issued, use burnt cork, bark, charcoal, lamp black, or light-colored mud.

CHAPTER 8

TRACKING

GENERAL

In all operations, you must be alert for signs of enemy activity. Such signs can often alert you to an enemy's presence and give your unit time to prepare for contact. The ability to track an enemy after he has broken contact also helps you regain contact with him.

TRACKER QUALITIES

Visual tracking is following the path of men or animals by the signs they leave, primarily on the ground or vegetation. Scent tracking is following men or animals by their smell.

Tracking is a precise art. You need a lot of practice to achieve and keep a high level of tracking skill. You should be familiar with the general techniques of tracking to enable you to detect the presence of a hidden enemy and to follow him, to find and avoid mines or booby-traps, and to give early warning of ambush.

With common sense and a degree of experience, you can track another person. However, you must develop the following traits and qualities:

- Be patient.
- Be able to move slowly and quietly, yet steadily, while detecting and interpreting signs.
- Avoid fast movement that may cause you to overlook signs, lose the trail, or blunder into an enemy unit.
- Be persistent and have the skill and desire to continue the mission even though signs are scarce or weather or terrain is unfavorable.
- Be determined and persistent when trying to find a trail that you have lost.

- Be observant and try to see things that are not obvious at first glance.
- Use your senses of smell and hearing to supplement your sight.
- Develop a feel for things that do not look right. It may help you regain a lost trail or discover additional signs.
- Know the enemy, his habits, equipment, and capability.

FUNDAMENTALS OF TRACKING

When tracking an enemy, you should build a picture of him in your mind. Ask yourself such questions as: How many persons am I following? How well are they trained? How are they equipped? Are they healthy? How is their morale? Do they know they are being followed?

To find the answer to such questions, use all available signs. A sign can be anything that shows you that a certain act took place at a particular place and time. For instance, a footprint tells a tracker that at a certain time a person walked on that spot.

The six fundamentals of tracking are:

- Displacement.
- Staining.
- Weathering.
- Littering.
- Camouflaging.
- Interpretation and/or immediate use intelligence.

Any sign that you find can be identified as one or more of the first five fundamentals.

In the sixth fundamental, you combine the first five and use all of them to form a picture of the enemy.

DISPLACEMENT

Displacement takes place when something is moved from its original position. An example is a footprint in soft, moist ground. The foot of the person that left the print displaced the soil, leaving an indentation in the ground. By studying the print, you can determine many facts. For example, a print that was left by a barefoot person or a person with worn or frayed footgear indicates that he may have poor equipment.

HOW TO ANALYZE FOOTPRINTS

Footprints show the following:

- The direction and rate of movement of a party.
- The number of persons in a party.
- Whether or not heavy loads are carried.
- The sex of the members of a party.
- Whether the members of a party know they are being followed.

If the footprints are deep and the pace is long, the party is moving rapidly. Very long strides and deep prints, with toe prints deeper than heel prints, indicate that the party is running. If the prints are deep, short, and widely spaced, with signs of scuffing or shuffling, a heavy load is probably being carried by the person who left the prints.

You can also determine a person's sex by studying the size and position of the footprints.

Women generally tend to be pigeon-toed, while men usually walk with their feet pointed straight ahead or slightly to the outside. Women's prints are usually smaller than men's, and their strides are usually shorter.

If a party knows that it is being followed, it may attempt to hide its tracks. Persons walking backward have a short, irregular stride. The prints have an unusually deep toe. The soil will be kicked in the direction of movement.

The last person walking in a group usually leaves the clearest footprints. Therefore, use his prints as the key set. Cut a stick the length of each key print and notch the stick to show the print width at the widest part of the sole. Study the angle of the key prints to determine the direction of march. Look for an identifying mark or feature on the prints, such as a worn or frayed part of the footwear. If the trail becomes vague or obliterated, or if the trail being followed merges with another, use the stick to help identify the key prints. That will help you stay on the trail of the group being followed.

Use the box method to count the number of persons in the group. There are two ways to use the box method—the stride as a unit of measure method and the 36-inch box method.

The stride as a unit of measure method is the more accurate of the two. Up to 18 persons can be counted using this method. Use it when the key prints can be determined. To use this method, identify a key print on a trail and draw a line from its heel across the trail.

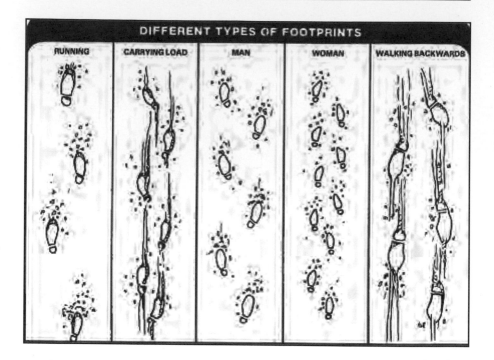

Then move forward to the key print of the opposite foot and draw a line through its instep. This should form a box with the edges of the trail forming two sides, and the drawn lines forming the other two sides. Next, count every print or partial print inside the box to determine the number of persons. Any person walking normally would have stepped in the box at least one time. Count the key prints as one.

To use the 36-inch box method, mark off a 30- to 36-inch cross section of a trail, count the prints in the box, then divide by two to determine the number of persons that used the trail. (Your M16 rifle is 39 inches long and may be used as a measuring device.)

OTHER SIGNS OF DISPLACEMENT

Footprints are only one example of displacement. Displacement occurs when anything is moved from its original position. Other examples are such things as foliage, moss, vines, sticks, or rocks that are moved from their original places; dew droplets brushed from leaves; stones and sticks that are turned over and show a different color underneath; and grass or other vegetation that is bent or broken in the direction of movement.

Bits of cloth may be torn from a uniform and left on thorns, snags, or the ground, and dirt from boots may make marks on the ground.

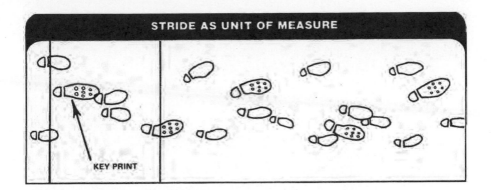

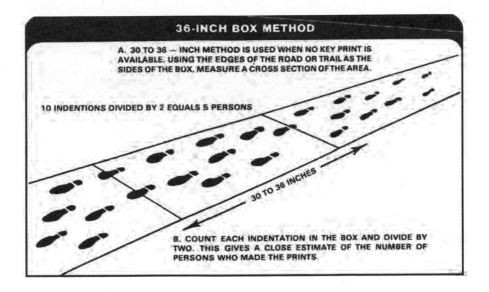

Another example of displacement is the movement of wild animals and birds that are flushed from their natural habitats. You may hear the cries of birds that are excited by strange movements. The movement of tall grass or brush on a windless day indicates that some-thing is moving the vegetation from its original position.

When you clear a trail by either breaking or cutting your way through heavy vegetation,you displace the vegetation. Displacement signs can be made while you stop to rest with heavy loads. The prints made by the equipment you carry can help to identify its type. When loads are set down at a rest halt or campsite, grass and twigs may be crushed. A sleeping man may also flatten the vegetation.

In most areas, there will be insects. Any changes in the normal life of these insects maybe a sign that someone has recently passed through the area. Bees that are stirred up, and holes that are covered by someone moving over them, or spider webs that are torn down are good clues.

If a person uses a stream to cover his trail, algae and water plants may be displaced in slippery footing or in places where he walks carelessly. Rocks may be displaced from their original position, or turned over to show a lighter or darker color on their opposite side. A person entering or leaving a stream may create slide marks, wet banks, or footprints, or he may scuff bark off roots or sticks. Normally, a person or animal will seek the path of least resistance. Therefore, when you search a stream for exit signs, look for open places on the banks or other places where it would be easy to leave the stream.

EXAMPLES OF DISPLACEMENT

TURNED OVER ROCKS AND STICKS

CRUSHED AND DISTURBED VEGETATION

SLIPMARKS AND WATER-FILLED FOOTPRINTS ON STREAM BANKS

STAINING

A good example of staining is the mark left by blood from a bleeding wound. Bloodstains often will be in the form of drops left by a wounded person. Blood signs are found on the ground and smeared on leaves or twigs.

You can determine the location of a wound on a man being followed by studying the bloodstains. If the blood seems to be dripping steadily, it probably came from a wound on his trunk. A wound in the lungs will deposit bloodstains that are pink, bubbly, frothy. A bloodstain deposited from a head wound will appear heavy, wet, and slimy, like gelatin. Abdominal wounds often mix blood with digestive juices so that the deposit will have an odor. The stains will be light in color.

Staining can also occur when a person walks over grass, stones, and shrubs with muddy boots. Thus, staining and displacement together may give evidence of movement and indicate the direction taken. Crushed leaves may stain rocky ground that is too hard for footprints.

Roots, stones, and vines may be stained by crushed leaves or berries when walked on. Yellow stains in snow may be urine marks left by personnel in the area.

In some cases, it may be hard to determine the difference between staining and displacement. Both terms can be applied to some signs. For example, water that has been muddied may indicate recent movement. The mud has been displaced and it is staining the water. Stones in streams may be stained by mud from boots. Algae can be displaced from stones in streams and can stain other stones or bark.

Water in footprints in swampy ground may be muddy if the tracks are recent. In time, however, the mud will settle and the water will clear. The clarity of the water can be used to estimate the age of the prints. Normally, the mud will clear in 1 hour. That will vary with terrain.

WEATHERING

Weather may either aid or hinder tracking. It affects signs in ways that help determine how old they are, but wind, snow, rain, and sunlight can also obliterate signs completely.

By studying the effects of weather on signs, you can determine the age of the sign.

For example, when bloodstains are fresh, they may be bright red. Air and sunlight will change the appearance of blood first to a deep ruby-red color, and then to a dark brown crust when the moisture evaporates. Scuff marks on trees or bushes darken with time. Sap oozes from fresh cuts on trees but it hardens when exposed to the air.

FOOTPRINTS

Footprints are greatly affected by weather. When a foot displaces soft, moist soil to form a print, the moisture holds the edges of the print intact and sharp. As sunlight and air dry the edges of the print, small particles that were held in place by the moisture fall into the print, making the edges appear rounded. Study this process carefully to estimate the age of a print. If particles are just beginning to fall into a print, it is probably fresh. If the edges of the print are dried and crusty, the prints are probably at least an hour old. The effects of weather will vary with the terrain, so this information is furnished as a guide only.

A light rain may round out the edges of a print. Try to remember when the last rain occurred in order to put prints into a proper time frame. A heavy rain may erase all signs.

Wind also affects prints. Besides drying out a print, the wind may blow litter, sticks, or leaves into it. Try to remember the wind activity in order to help determine the age of a print. For example, you may think, "It is calm now, but the wind blew hard an hour ago. These prints have litter blown into them, so they must be over an hour old." You must be sure, however, that the litter was blown into the prints, and was not crushed into them when the prints were made.

Trails leaving streams may appear to be weathered by rain because of water running into the footprints from wet clothing or equipment. This is particularly true if a party leaves a stream in a file. From this formation, each person drips water into the prints. A wet trail slowly fading into a dry trail indicates that the trail is fresh.

WIND, SOUNDS, AND ODORS

Wind affects sounds and odors. If the wind is blowing from the direction of a trail you are following, sounds and odors are carried to you. If the wind is blowing in the same direction as the trail you are following, you must be cautious as the wind will carry your sounds toward the enemy. To find the wind direction, drop a handful of dry dirt or grass from shoulder height.

To help you decide where a sound is coming from, cup your hands behind your ears and slowly turn. When the sound is loudest, you are probably facing the origin of sound. When moving, try to keep the wind in your face.

SUN

You must also consider the effects of the sun. It is hard to look or aim directly into the sun. If possible, keep the sun at your back.

LITTERING

Poorly trained units may leave trails of litter as they move. Gum or candy wrappers, ration cans, cigarette butts, remains of fires, or human feces are unmistakable signs of recent movement.

Weather affects litter. Rain may flatten or wash litter away, or turn paper into pulp. Winds may blow litter away from its original location. Ration cans exposed to weather will rust. They first rust at the exposed edge where they were opened. Rust then moves in toward the center. Use your memory to determine the age of litter. The last rain or strong wind can be the basis of a time frame.

CAMOUFLAGE

If a party knows that you are tracking it, it will probably use camouflage to conceal its movement and to slow and confuse you. Doing so, however, will slow it down. Walking backward, brushing out trails, and moving over rocky ground or through streams are examples of camouflage that can be used to confuse you.

The party may move on hard surfaced, frequently traveled roads or try to merge with traveling civilians. Examine such routes with extreme care, because a well-defined approach that leads to the enemy will probably be mined, ambushed, or covered by snipers.

The party may try to avoid leaving a trail. Its members may wrap rags around their boots, or wear soft-soled shoes to make the edges of their footprints rounder and less distinct. The party may exit a stream in column or line to reduce the chance of leaving a well-defined exit.

If the party walks backward to leave a confusing trail, the footprints will be deepened at the toe, and the soil will be scuffed or dragged in the direction of movement.

If a trail leads across rocky or hard ground, try to work around that ground to pick up the exit trail. This process works in streams as well. On rocky ground, moss or lichens growing on the stones could be displaced by even the most careful evader. If you lose the trail, return to the last visible sign. From there, head in the direction of the party's movement. Move in ever-widening circles until you find some signs to follow.

INTERPRETATION/IMMEDIATE USE INTELLIGENCE

When reporting, do not report your interpretations as facts. Report that you have seen signs of certain things, not that those things actually exist.

Report all information quickly. The term "immediate use intelligence" includes information of the enemy that can be put to use at once to gain surprise, to keep the enemy off balance, or to keep him from escaping an area .A commander has many sources of intelligence. He puts the information from those sources together to help determine where an enemy is, what he may be planning, and where he may be going.

Information you report gives your leader definite information on which he can act at once. For example, you may report that your leader is 30 minutes behind an enemy unit, that the enemy is moving north, and that he is now at a certain place. That gives your leader information on which he can act at once. He could then have you keep on tracking and move another unit to attack the enemy. If a trail is found that has signs of recent enemy activity, your leader can set up an ambush on it.

TRACKING TEAMS

Your unit may form tracking teams. The lead team of a moving unit can be a tracking team, or a separate unit may be a tracking team. There are many ways to organize such teams, and they can be any size. There should, however, be a leader, one or more trackers, and security for the trackers. A typical organization has three trackers, three security men, and a team leader with a radiotelephone operator (RATELO).

When a team is moving, the best tracker should be in the lead, followed by his security. The two other trackers should be on the flanks, each one followed and overmatched by his security. The leader should be where he can best control the team. The RATELO should be with the leader.

COUNTERTRACKING

In addition to knowing how to track, you must know how to counter an enemy tracker's efforts to track you. Some countertracking techniques are discussed in the following paragraphs:

- While moving from close terrain to open terrain, walk past a big tree (30 cm [12 in] in diameter or larger)toward the open area for three to five paces. Then walk backward to the forward side of the tree and make a 90-degree change of direction, passing the tree on its forward side. Step carefully and leave as little sign as possible. If this is not the direction that you want to go, change direction again about 50 meters away using the same technique. The purpose of this is to draw the enemy tracker into the open area where it is harder for him to track. That also exposes him and causes him to search the wrong area.

- When approaching a trail (about 100 meters from it), change your direction of movement and approach it at a 45-degree angle. When arriving at the trail, move along it for about 20 to 30 meters. Leave several signs of your presence. Then walk backward along the trail to the point where you join edit. At that point, cross the trail and leave no sign of your leaving it. Then move about 100 meters at an angle of 45 degrees, but this time on the other side of the trail and in the reverse of your approach. When changing direction back to your original line of march, the big tree technique can be used. The purpose of this technique is to draw the enemy tracker along the easier trail. You have, by changing direction before reaching the trail, indicated that the trail is your new line of march.

- To leave a false trail and to get an enemy tracker to look in the wrong direction, walk backward over soft ground. Continue this deception for about 20 to 30 meters or until you are on hard ground. Use this technique when leaving a stream. To further confuse the enemy tracker, use this technique several times before actually leaving the stream.

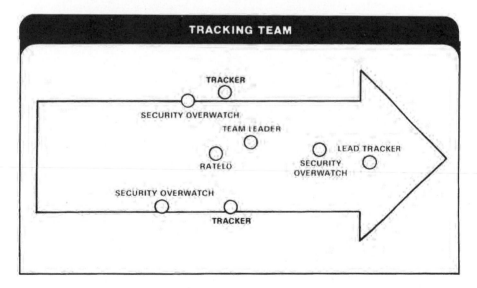

TRACKING TEAM

TRACKER

SECURITY OVERWATCH

TEAM LEADER

LEAD TRACKER

RATELO

SECURITY
OVERWATCH

SECURITY OVERWATCH

TRACKER

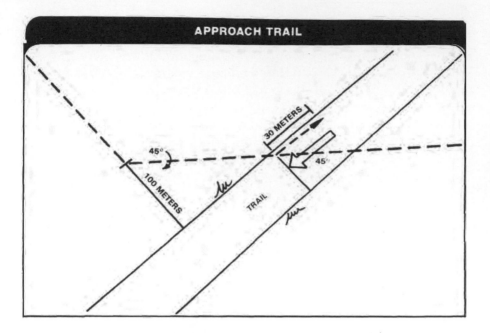

- When moving toward a stream, change direction about 100 meters before reaching the stream and approach it at a 45-degree angle. Enter the stream and proceed down it for at least 20 to 30 meters. Then move back upstream and leave the stream in your initial direction. Changing direction before entering the stream may confuse the enemy tracker. When he enters the stream, he should follow the false trail until the trail is lost. That will put him well away from you.
- When your direction of movement parallels a stream, use the stream to deceive an enemy tracker. Some tactics that will help elude a tracker areas follows:
  o Stay in the stream for 100 to 200 meters.
  o Stay in the center of the stream and in deep water.
  o Watch for rocks or roots near the banks that are not covered with moss or vegetation and leave the stream at that point.
  o Walk out backward on soft ground.
  o Walk up a small, vegetation-covered tributary and exit from it.
- When being tracked by an enemy tracker, the-best bet is to either try to outdistance him or to double back and ambush him.

FALSE TRAIL LEAVING STREAM

20-30 METERS

HARD GROUND

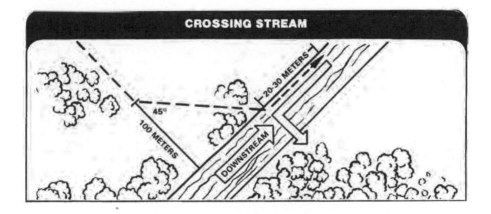

CROSSING STREAM

45°

100 METERS

20-30 METERS

DOWNSTREAM

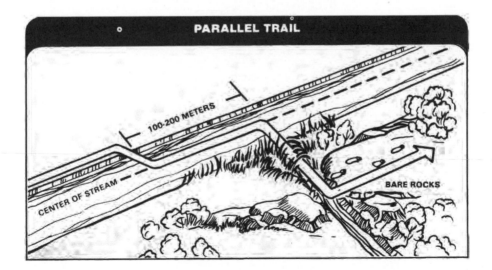

PARALLEL TRAIL

100-200 METERS

CENTER OF STREAM

BARE ROCKS

CHAPTER 9

MOVEMENT

GENERAL

Normally, you will spend more time moving than fighting. You must use proper movement techniques to avoid contact with the enemy when you are not prepared for contact.

The fundamentals of movement discussed in this chapter provide techniques that all soldiers should learn. These techniques should be practiced until they become second nature.

MOVEMENT TECHNIQUES

Your unit's ability to move depends on your movement skills and those of your fellow soldiers. Use the following techniques to avoid being seen or heard by the enemy:

- Camouflage yourself and your equipment.
- Tape your dog tags together and to the chain so they cannot slide or rattle. Tape or pad the parts of your weapon and equipment that rattle or are so loose that they may snag (the tape or padding must not interfere with the operation of the weapon or equipment).
- Jump up and down and listen for rattles.
- Wear soft, well-fitting clothes.
- Do not carry unnecessary equipment. Move from covered position to revered position (taking no longer than 3 to 5 seconds between positions).
- Stop, look, and listen before moving. Look for your next position before leaving a position.
- Look for covered and concealed routes on which to move.
- Change direction slightly from time to time when moving through tall grass.

- Stop, look, and listen when birds or animals are alarmed (the enemy maybe nearby).
- Use battlefield noises, such as weapon noises, to conceal movement noises.
- Cross roads and trails at places that have the most cover and concealment (large culverts, low spots, curves, or bridges).
- Avoid steep slopes and places with loose dirt or stones.
- Avoid cleared, open areas and tops of hills and ridges.

METHODS OF MOVEMENT

In addition to walking, you may move in one of three other methods—low crawl, high crawl, or rush.

The low crawl gives you the lowest silhouette. Use it to cross places where the concealment is very low and enemy fire or observation prevents you from getting up. Keep your body flat against the ground. With your firing hand, grasp your weapon sling at the upper sling—swivel. Let the front hand guard rest on your forearm (keeping the muzzle off the ground), and let the weapon butt drag on the ground.

To move, push your arms forward and pull your firing side leg forward. Then pull with your arms and push with your leg. Continue this throughout the move.

The high crawl lets you move faster than the low crawl and still gives you a low silhouette. Use this crawl when there is good concealment but enemy fire prevents you from getting up. Keep your body off the ground and resting on your forearms and lower legs. Cradle your weapon in your arms and keep its muzzle off the ground. Keep your knees well behind your buttocks so your body will stay low.

To move, alternately advance your right elbow and left knee, then your left elbow and right knee.

The rush is the fastest way to move from one position to another. Each rush should last from 3 to 5 seconds. The rushes are kept short to keep enemy machine gunners or riflemen from tracking you. However, do not stop and hit the ground in the open just because 5 seconds have passed. Always try to hit the ground behind some cover. Before moving, pick out your next covered and concealed position and the best route to it.

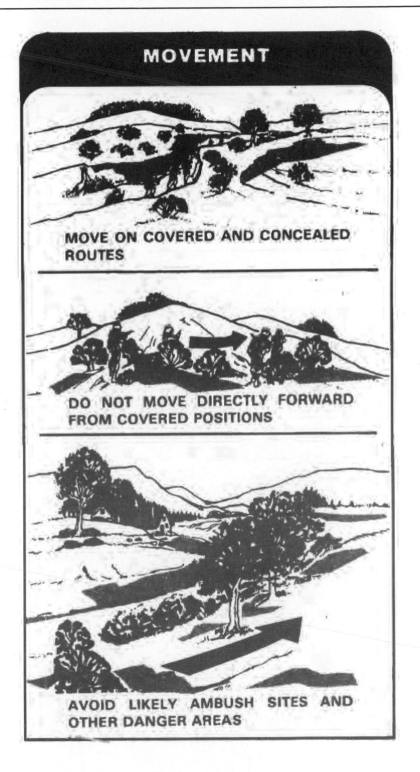

MOVEMENT

MOVE ON COVERED AND CONCEALED ROUTES

DO NOT MOVE DIRECTLY FORWARD FROM COVERED POSITIONS

AVOID LIKELY AMBUSH SITES AND OTHER DANGER AREAS

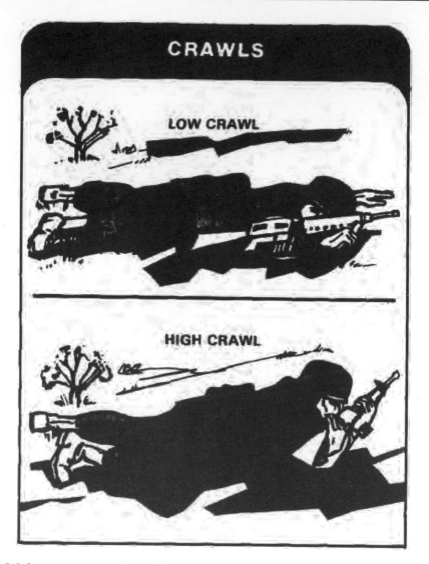

Make your move from the prone position as follows:

- Slowly raise your head and pick your next position and the route to it.
- Slowly lower your head.
- Draw your arms into your body (keeping your elbows in).
- Pull your right leg forward.
- Raise your body by straightening your arms.
- Get up quickly.
- Run to the next position.

When you are ready to stop moving, do the following:

- Plant both of your feet.
- Drop to your knees (at the same time slide a hand to the butt of your rifle).
- Fall forward, breaking the fall with the butt of the rifle.
- Go to a prone firing position.

If you have been firing from one position for some time, the enemy may have spotted you and may be waiting for you to come up from behind cover. So, before rushing forward, roll or crawl a short distance from your position. By coming up from another spot, you may fool an enemy who is aiming at one spot, waiting for you to rise.

When the route to your next position is through an open area, rush by zigzagging. If necessary, hit the ground, roll right or left, then rush again.

MOVING WITH STEALTH

Moving with stealth means moving quietly, slowly, and carefully. This requires great patience.

To move with stealth, use the following techniques:

- Hold your rifle at port arms (ready position).
- Make your footing sure and solid by keeping your body's weight on the foot on the ground while stepping.
- Raise the moving leg high to clear brush or grass.
- Gently let the moving foot down toe first, with your body's weight on the rear leg.
- Lower the heel of the moving foot after the toe is in a solid place.
- Shift your body's weight and balance to the forward foot before moving the rear foot.
- Take short steps to help maintain balance.

REACTING TO AERIAL FLARES

At night, and when moving through dense vegetation, avoid making noise. Hold your weapon with one hand, and keep the other hand forward, feeling for obstructions.

When going into a prone position, use the following techniques:

- Hold your rifle with one hand and crouch slowly.
- Feel for the ground with your freehand to make sure it is clear of mines, tripwires, and other hazards.
- Lower your knees, one at a time, until your body's weight is on both knees and your free hand.

- Shift your weight to your free hand and opposite knee.
- Raise your free leg up and back, and lower it gently to that side.
- Move the other leg into position the same way.
- Roll quietly into a prone position.
  Use the following techniques when crawling:
- Crawl on your hands and knees. Hold your rifle in your firing hand.
- Use your nonfiring hand to feel for and make clear spots for your hands and knees to move to.
- Move your hands and knees to those spots, and put them down softly.

IMMEDIATE ACTIONS WHILE MOVING

This section furnishes guidance for the immediate actions you should take when reacting to enemy indirect fire and flares.

Reacting to Indirect Fire

If you come under indirect fire while moving, quickly look to your leader for orders.

He will either tell you to run out of the impact area in a certain direction or will tell you to follow him. If you cannot see your leader, but can see other team members, follow them. If alone, or if you cannot see your leader or the other team members, run out of the area in a direction away from the incoming fire.

It is hard to move quickly on rough terrain, but the terrain may provide good cover. In such terrain, it may be best to take cover and wait for flares to burn out. After they burnout, move out of the area quickly.

Reacting to Ground Flares

The enemy puts out ground flares as warning devices. He sets them off himself or attaches tripwires to them for you to trip on and set them off. He usually puts the flares in places he can watch.

If you are caught in the light of a ground flare, move quickly out of the lighted area. The enemy will know where the ground flare is and will be ready to fire into that area. Move well away from the lighted area. While moving out of the area, look for other team members. Try to follow or join them to keep the team together.

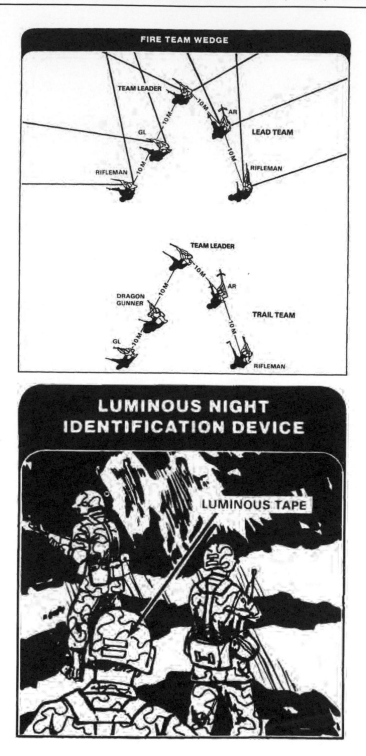

Reacting to Aerial Flares

The enemy uses aerial flares to light up vital areas. They can be set off like ground flares; fired from hand projectors, grenade launchers, mortars, and artillery; or dropped from aircraft.

If you hear the firing of an aerial flare while you are moving, hit the ground (behind cover if possible) while the flare is rising and before it bursts and illuminates.

If moving where it is easy to blend with the background (such as in a forest) and you are caught in the light of an aerial flare, freeze in place until the flare burns out.

If you are caught in the light of an aerial flare while moving in an open area, immediately crouch low or lie down.

If you are crossing an obstacle, such as a barbed-wire fence or a wall, and get caught in the light of an aerial flare, crouch low and stay down until the flare burns out.

The sudden light of a bursting flare may temporarily blind both you and the enemy. When the enemy uses a flare to spot you, he spoils his own night vision. To protect your night vision, close one eye while the flare is burning. When the flare burns out, the eye that was closed will still have its night vision.

MOVING WITHIN A TEAM

You will usually move as a member of a team. Small teams, such as infantry fire teams, normally move in a wedge formation. Each soldier in the team has a set position in the wedge, determined by the type weapon he carries. That position, however, may be changed by the team leader to meet the situation. The normal distance between soldiers is 10 meters.

You may have to make a temporary change in the wedge formation when moving through close terrain. The soldiers in the sides of the wedge close into a single file when moving in thick brush or through a narrow pass. After passing through such an area, they should spread out, again forming the wedge. You should not wait for orders to change the formation or the interval. You should change automatically and stay in visual contact with the other team members and the team leader.

FIRE TEAM WEDGE

The team leader leads by setting the example. His standing order is, FOLLOW ME AND DO AS I DO. When he moves to the left, you should move to the left. When he gets down, you should get down. When he fires, you should fire.

When visibility is limited, control during movement may become difficult. Two l-inch horizontal strips of luminous tape, sewn directly on the rear of the helmet camouflage band with a l-inch space between them, are a device for night identification.

Night identification for your patrol cap could be two l-inch by 1/2-inch strips of luminous tape sewn vertically, directly on the rear of the cap. They should be centered, with the bottom edge of each tape even with the bottom edge of the cap and with a l-inch space between the two tapes.

FIRE AND MOVEMENT

When a unit makes contact with the enemy, it normally starts firing at and moving toward the enemy. Sometimes the unit may move away from the enemy. That technique is called fire and movement. It is conducted either to close with and destroy the enemy, or to move away from the enemy so as to break contact with him.

The firing and moving take place at the same time. There is a fire element and a movement element. These elements may be single soldiers, buddy teams, fire teams, or squads. Regardless of the size of the elements, the action is still fire and movement.

The fire element covers the move of the movement element by firing at the enemy. This helps keep the enemy from firing back at the movement element.

The movement element moves either to close with the enemy or to reach a better position from which to fire at him. The movement element should not move until the fire element is firing.

Depending on the distance to the enemy position and on the available cover, the fire element and the movement element switch roles as needed to keep moving.

Before the movement element moves beyond the supporting range of the fire element (the distance within which the weapons of the fire element can fire and support the movement element), it should take a position from which it can fire at the enemy. The movement element then becomes the next fire element and the fire element becomes the next movement element.

If your team makes contact, your team leader should tell you to fire or to move. He should also tell you where to fire from, what to fire at, or where to move to. When moving, use the low crawl, high crawl, or rush.

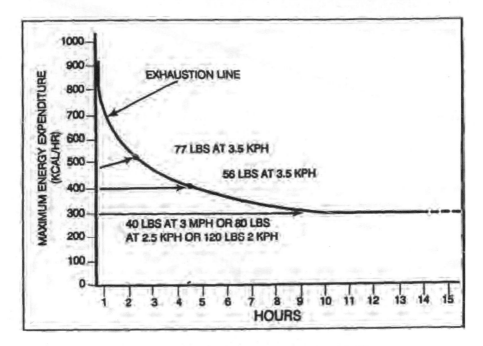

Figure 9-8: Work Rate and Energy Expenditure

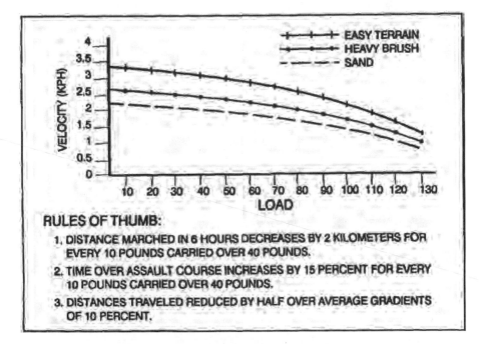

Figure 9-9: March speeds

FOOT MARCH LOADS

The fighting load for a conditioned soldier should not exceed 48 pounds and the approach march load should not exceed 72 pounds. These load weights include all clothing and equipment that are worn and carried.

   a. A soldier's ability to react to the enemy is reduced by the burden of his load. Load carrying causes fatigue and lack of agility, placing soldiers at a disadvantage when rapid reaction to the enemy is required. For example, the time a soldier needs to complete an obstacle course is increased from 10 to 15 percent, depending on the configuration of the load, for every 10 pounds of equipment carried. It is likely that a soldier's agility in the assault will be degraded similarly.
   b. Speed of movement is as important a factor in causing exhaustion as the weight of the load carried. The chart at Figure 5-1 shows the length of time that work rates can be sustained before soldiers become exhausted and energy expenditure rates for march speeds and loads. A burst rate of energy expenditure of 900 to 1,000 calories per hour can only be sustained for 6 to 10 minutes. Fighting loads must be light so that the bursts of energy available to a soldier are used to move and to fight, rather than to carry more than the minimum fighting equipment.
   c. When carrying loads during approach marches, a soldier's speed can cause a rate-of-energy expenditure of over 300 calories per hour and can erode the reserves of energy needed upon enemy contact. March speeds must be reduced when loads are heavier to stay within reasonable energy expenditure rates. Carrying awkward loads and heavy handheld items causes further degradation of march speed and agility. The distance marched in six hours decreases by about 2 km for every 10 pounds carried over 40 pounds.

Battlefield stress decreases the ability of soldiers to carry their loads. Fear burns up the glycogen in the muscles required to perform physical tasks. This wartime factor is often overlooked in peacetime, but the commander must consider such a factor when establishing the load for each soldier. However, applying strong leadership to produce well-trained, highly motivated soldiers can lessen some of the effects of stress.

As the modern battlefield becomes more sophisticated, potential enemies develop better protected equipment, which could be presented as fleeting targets. Unless technological breakthroughs occur, increasingly heavy munitions and new types of target acquisition and communications equipment will be required by frontline soldiers to defeat the enemy.

    a. In the future, the foot soldier's load can be decreased only by sending him into battle inadequately equipped or by providing some means of load-handling equipment to help him carry required equipment.

WEIGHT IN POUNDS *(every ounce counts)*	
BDU 3.8	TROUSERS, WET WEATHER 1.2
DRAWERS, COTTON 1	RATION, MRE 1.3
HANDKERCHIEF................................. .1	BAG, WATERPROOF 8
SOCK, CUSHION SOLE 2	PAD, SLEEPING 1.3
UNDERSHIRT, COTTON 3	3 SHELTER HALF, POLES,
TOWEL... .2	PEGS, AND ROPE 4.5
CANTEEN, 1 qt w/WATER 2.8	CARRIER, SLEEPING BAG 4
CANTEEN, 2 qt w/WATER 4.8	BOOTS, COMBAT LEATHER 3.3
LINER, PONCHO 1.6	JACKET, FIELD 3.3
MESS KIT... 2.8	BAG, DUFFLE 3.5
GLOVES, BARBED WIRE 4	CAP, BDU 3
PARKA, WET WEATHER 1.2	CASE, SLEEPING BAG 1.5
PONCHO, NYLON 1.3	LINER, FIELD JACKET 7
SHIRT, SLEEPING 7	OVERSHOES 4.2
SCARF, WOOL 4	TELEPHONE, TA-1 .... 1.5
SLEEPING BAG 7.1	E-TOOL, w/CARRIER....................... 2.5
BELT, TROUSERS 2	ALICE, MEDIUM, w/FRAME............. 6.3
HELMET, BALLISTIC 3.4	ALICE, LARGE, w/FRAME............... 6.6
BELT, PISTOL, w/SUSPENDERS	AN/PRC-77, w/BATTERY................ 20.8
AND FIRST-AID POUCH 1.6	M60 SPARE BARREL w/BAG 8.0
TOILET ARTICLES 2.0	60-mm MORTAR, M225................. 14.4
WEAPONS:	60-mm SIGHT, M64........................ 2.5
M16.. 1.6	60-mm BASEPLATE, M-7 .... 14.4
M203...................................... 10.0	60-mm BIPOD................................ 13.2
M60 MG 23.3	81-mm MORTAR, M29.................. 30.0
M249 SAW 15.2	81-mm SIGHT, M53....................... 6.0
AMMUNITION:	81-mm NIGHTLIGHT....................... 2.0
5.56 w/MAG (30 rds)..................... .9	81-mm BASEPLATE....................... 25.0
7.62 LKD (100 rds) 7.0	81-mm BIPOD 40.0
40-mm (ALL TYPES)...................... .5	BINOCULARS.................................. 3.2
5.56 LKD (200 rds) 7.6	FLASHLIGHT, w/BATTERY............... .8
GRENADE, FRAGMENTATION.. 1.0	COMPASS, M2................................. .3
GRENADE, SMOKE 1.0	DRAGON TRACKER........................ 8.1
RD, 60-mm MORTAR, HE 3.5	DRAGON NIGHTSIGHT.................. 34.0
RD, 81-mm MORTAR, HE 9.3	AN/PVS-5 NVG................................ 1.9
LAW.. 5.2	AN/PVS-4 NVD................................ 3.9
MINE, M21 .... 18.0	PISTOL, CAL .45............................. 2.5
CLAYMORE, M18 .... 5.0	PROTECTIVE MASK, w/DECON KIT 3.0
DRAGON, MSL 25.3	
AT4... 14.0	
FLARE, TRIP............................. 1.0	
BAYONET, w/SCABBARD 1.3	
CASE, SMALL-ARMS 9	

Figure 9-10: Weights of selected items

b. Unless part of the load is removed from the soldier's back and carried elsewhere, all individual load weights are too heavy. Even if rucksacks are removed, key teams on the battlefield cannot fulfill their roles unless they carry excessively heavy loads. Soldiers who must carry heavy loads restrict the mobility of their units.

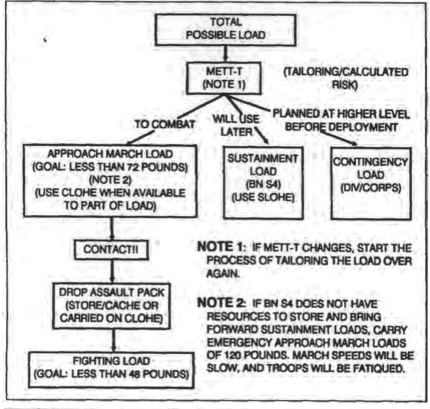

CONFLICT/ ENVIRONMENT	CLOHE (COMPANY LEVEL)	SLOHE (BATTALION LEVEL)
LOW INTENSITY		
TEMPERATE CLIMATE	1,850 LB	4,800 LB/400 CUBIC FT
COLD WET CLIMATE	1,850 LB	7,000 LB/600 CUBIC FT
MEDIUM INTENSITY		
TEMPERATE CLIMATE	2,250 LB	6,350 LB/500 CUBIC FT
COLD WET CLIMATE	2,250 LB	8,550 LB/700 CUBIC FT

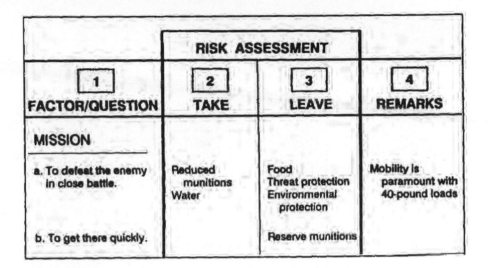

RISK ASSESSMENT			
1 FACTOR/QUESTION	**2** TAKE	**3** LEAVE	**4** REMARKS
MISSION a. To defeat the enemy in close battle. b. To get there quickly.	Reduced munitions Water	Food Threat protection Environmental protection Reserve munitions	Mobility is paramount with 40-pound loads

	RISK ASSESSMENT		
1 **FACTOR/QUESTION**	**2** **TAKE**	**3** **LEAVE**	**4** **REMARKS**
c. To sustain stealth operations independent of resupply.	Water Food Environmental protection Reduced munitions Camouflage	Reserve munitions Threat protection	Maximum loads depend on speed/distance for dynamic operations.
d. To carry maximum combat power.	Munitions Water Threat protection Limited environmental protection	Food	Maximum loads depend on speed/distance for dynamic operations.
e. For static operations.	Basic load and reserve of ammunition Barrier materiel Maximum threat protection Some comfort items to achieve quality rest periods Water Food		
RESUPPLY			
a. Reliability.	Less amounts of all classes of supply		Best solution is for the commander to control his own immediate resupply transport resources.
b. On call.		Reserve munitions	

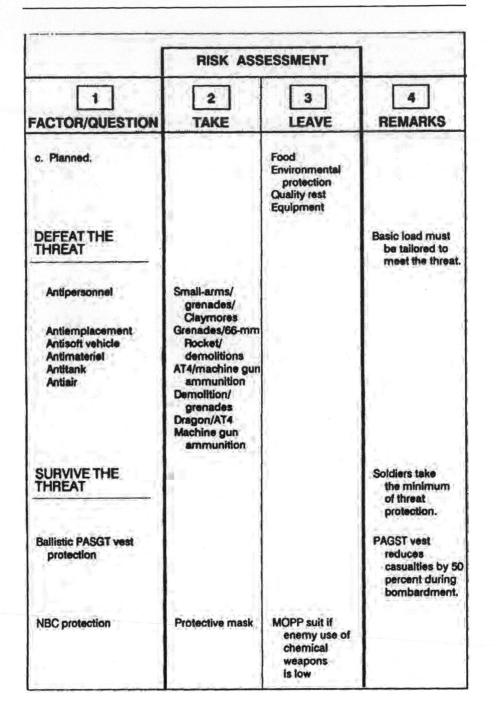

| | RISK ASSESSMENT | | |
1 FACTOR/QUESTION	**2** TAKE	**3** LEAVE	**4** REMARKS
c. Planned.		Food Environmental protection Quality rest Equipment	
DEFEAT THE THREAT			Basic load must be tailored to meet the threat.

Antipersonnel	Small-arms/
	    grenades/
	    Claymores		

Antiemplacement
Antisoft vehicle
Antimateriel
Antitank
Antiair
	Grenades/66-mm
	    Rocket/
	    demolitions
	AT4/machine gun
	    ammunition
	Demolition/
	    grenades
	Dragon/AT4
	Machine gun
	    ammunition		

SURVIVE THE THREAT			Soldiers take the minimum of threat protection.
Ballistic PASGT vest protection			PAGST vest reduces casualties by 50 percent during bombardment.
NBC protection	Protective mask	MOPP suit if enemy use of chemical weapons is low	

	RISK ASSESSMENT		
1 FACTOR/QUESTION	**2** TAKE	**3** LEAVE	**4** REMARKS
Electronic Warfare	VINSON		Secure communi-cations probably not viable below bde/bn level in light units unless COMSEC is of high priority to achieve mission.
TERRAIN			
Flat, improved road			Terrain may cause an increase of time required to conduct march; resupply cross country may be difficult.
Cross country	Water consumption increased		
Hills, improved road			
WEATHER			Energy must be maintained to fight by control of loads/march speeds.
a. Environmental Survival:			

	RISK ASSESSMENT		
1 FACTOR/QUESTION	**2** TAKE	**3** LEAVE	**4** REMARKS
Exposure	Poncho Extra clothing Limited number of sleeping bags		Work rates should be reduced.
Heat exhaustion	Water	Threat protection	
Disease	Water purification tablets Mosquito nets		When in combat, men with excess fat can survive off natural reserves.
b. Sustenance	High-caloric food		Average of four hours quality sleep each day.
c. Quality rest	Sleeping bags/ pads		

CHAPTER 10

FIELD-EXPEDIENT DIRECTION FINDING

In a survival situation, you will be extremely fortunate if you happen to have a map and compass. If you do have these two pieces of equipment, you will most likely be able to move toward help. If you are not proficient in using a map and compass, you must take the steps to gain this skill.

There are several methods by which you can determine direction by using the sun and the stars. These methods, however, will give you only a general direction. You can come up with a more nearly true direction if you know the terrain of the territory or country.

You must learn all you can about the terrain of the country or territory to which you or your unit may be sent, especially any prominent features or landmarks. This knowledge of the terrain together with using the methods explained below will let you come up with fairly true directions to help you navigate.

USING THE SUN AND SHADOWS

The earth's relationship to the sun can help you to determine direction on earth. The sun always rises in the east and sets in the west, but not exactly due east or due west. There is also some seasonal variation. In the northern hemisphere, the sun will be due south when at its highest point in the sky, or when an object casts no appreciable shadow. In the southern hemisphere, this same noonday sun will mark due north. In the northern hemisphere, shadows will move clockwise. Shadows will move counterclockwise in the southern hemisphere. With practice, you can use shadows to determine both direction and time of day. The shadow methods used for direction finding are the shadow-tip and watch methods.

Shadow-Tip Methods

In the first shadow-tip method, find a straight stick 1 meter long, and a level spot free of brush on which the stick will cast a definite shadow. This method is simple and accurate and consists of four steps:

Step 1. Place the stick or branch into the ground at a level spot where it will east a distinctive shadow. Mark the shadow's tip with a stone, twig, or other means. This first shadow mark is always west —everywhere on earth.

Step 2. Wait 10 to 15 minutes until the shadow tip moves a few centimeters. Mark the shadow tip's new position in the same way as the first.

Step 3. Draw a straight line through the two marks to obtain an approximate east-west line.

Step 4. Stand with the first mark (west) to your left and the second mark to your right-you are now facing north. This fact is true everywhere on earth.

An alternate method is more accurate but requires more time. Set up your shadow stick and mark the first shadow in the morning. Use a piece of string to draw a clean arc through this mark and around the stick. At midday, the shadow will shrink and disappear. In the afternoon, it will lengthen again and at the point where it touches the arc, make a second mark. Draw a line through the two marks to get an accurate east-west line (see Figure 10-1).

The Watch Method

You can also determine direction using a common or analog watch—one that has hands. The direction will be accurate if you are using true local time, without any changes for daylight savings time. Remember, the further you are from the equator, the more accurate this method will be. If you only have a digital watch, you can overcome this obstacle. Quickly draw a watch on a circle of paper with the correct time on it and use it to determine your direction at that time.

In the northern hemisphere, hold the watch horizontal and point the hour hand at the sun. Bisect the angle between the hour hand and the12 o'clock mark to get the north-south line (Figure 10-2). If there is any doubt as to which end of the line is north, remember that the sun rises in the east, sets in the west, and is due south at noon. The sun is in the east before noon and in the west after noon.

1 Mark the shadow's tip.

2 Mark the new position and draw a line through the two marks.

3 Stand with the first mark to your left and the second mark to your right—you are now facing north.

Figure 10-1: Shadow-tip method

Note: If your watch is set on daylight savings time, use the midway point between the hour hand and 1 o'clock to determine the north-south line.

In the southern hemisphere, point the watch's 12 o'clock mark toward the sun and a midpoint halfway between 12 and the hour hand will give you the north-south line (Figure 10-2).

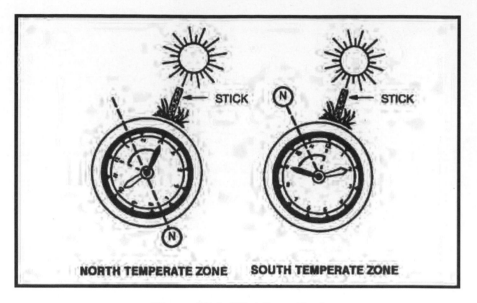

Figure 10-2: Watch method

USING THE MOON

Because the moon has no light of its own, we can only see it when it reflects the sun's light. As it orbits the earth on its 28-day circuit, the shape of the reflected light varies according to its position. We say there is a new moon or no moon when it is on the opposite side of the earth from the sun. Then, as it moves away from the earth's shadow, it begins to reflect light from its right side and waxes to become a full moon before waning, or losing shape, to appear as a sliver on the left side. You can use this information to identify direction.

If the moon rises before the sun has set, the illuminated side will be the west. If the moon rises after midnight, the illuminated side will be the east. This obvious discovery provides us with a rough east-west reference during the night.

USING THE STARS

Your location in the Northern or Southern Hemisphere determines which constellation you use to determine your north or south direction.

The Northern Sky

The main constellations to learn are the Ursa Major, also known as the Big Dipper or the Plow, and Cassiopeia (Figure 10-3). Neither of these constellations ever sets. They are always visible on a clear night. Use them to locate Polaris, also known as the polestar or the North Star. The North Star forms part of the Little Dipper handle and can be confused with the Big Dipper. Prevent confusion by using both the Big Dipper and Cassiopeia together. The Big Dipper and Cassiopeia are always directly opposite each. other and rotate counterclockwise around Polaris, with Polaris in the center. The Big Dipper is a seven star constellation in the shape of a dipper. The two stars forming the outer lip of this dipper are the "pointer stars" because they point to the North Star. Mentally draw a line from the outer bottom star to the outer top star of the Big Dipper's bucket. Extend this line about five times the distance between the pointer stars. You will find the North Star along this line.

Cassiopeia has five stars that form a shape like a "W" on its side. The North Star is straight out from Cassiopeia's center star.

After locating the North Star, locate the North Pole or true north by drawing an imaginary line directly to the earth.

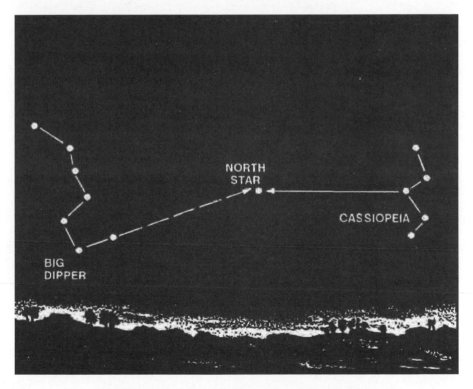

Figure 10-3: The Big Dipper and Cassiopeia

The Southern Sky

Because there is no star bright enough to be easily recognized near the south celestial pole, a constellation known as the Southern Cross is used as a signpost to the South (Figure 10-4). The Southern Cross or Crux has five stars. Its four brightest stars form a cross that tilts to one side. The two stars that make up the cross's long axis are the pointer stars. To determine south, imagine a distance five times the distance between these stars and the point where this imaginary line ends is in the general direction of south. Look down to the horizon from this imaginary point and select a landmark to steer by. In a static survival situation, you can fix this location in daylight if you drive stakes in the ground at night to point the way.

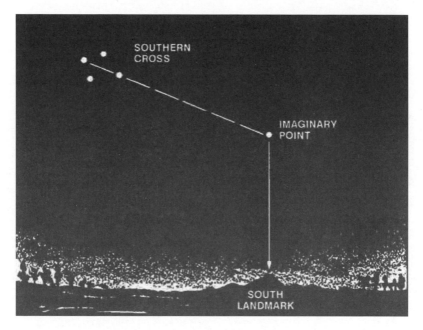

Figure 10-4: The Southern Cross

MAKING IMPROVISED COMPASSES

You can construct improvised compasses using a piece of ferrous metal that can be needle shaped or a flat double-edged razor blade and a piece of nonmetallic string or long hair from which to suspend it. You can magnetize or polarize the metal by slowly stroking it in one direction on a piece of silk or carefully through your hair using deliberate strokes. You can also polarize metal by stroking it repeatedly at one end with a magnet. Always rub in one direction only. If you have a battery and some electric wire, you can polarize the metal electrically. The wire should be insulated. If not insulated, wrap the metal object in a single, thin strip of paper to prevent contact. The battery must be a minimum of 2 volts. Form a coil with the electric wire and touch its ends to the battery's terminals. Repeatedly insert one end of the metal object in and out of the coil. The needle will become an electromagnet. When suspended from a piece of nonmetallic string, or floated on a small piece of wood in water, it will align itself with a north-south line.

You can construct a more elaborate improvised compass using a sewing needle or thin metallic object, a nonmetallic container (for example, a plastic dip container), its lid with the center cut out and waterproofed, and the silver tip from a pen. To construct this compass, take an ordinary sewing needle and break in half. One half will form your direction pointer and the other will act as the pivot point. Push the portion used as the pivot point through the bottom center of your container; this portion should be flush on the bottom and not interfere with the lid. Attach the center of the other portion (the pointer) of the needle on the pen's silver tip using glue, tree sap, or melted plastic. Magnetize one end of the pointer and rest it on the pivot point.

OTHER MEANS OF DETERMINING DIRECTION

The old saying about using moss on a tree to indicate north is not accurate because moss grows completely around some trees. Actually, growth is lusher on the side of the tree facing the south in the Northern Hemisphere and vice versa in the Southern Hemisphere. If there are several felled trees around for comparison, look at the stumps. Growth is more vigorous on the side toward the equator and the tree growth rings will be more widely spaced. On the other hand, the tree growth rings will be closer together on the side toward the poles. Wind direction may be helpful in some instances where there are prevailing directions and you know what they are.

Recognizing the differences between vegetation and moisture patterns on north- and south-facing slopes can aid in determining direction. In the northern hemisphere, north-facing slopes receive less sun than south-facing slopes and are therefore cooler and damper. In the summer, north-facing slopes retain patches of snow. In the winter, the trees and open areas on south-facing slopes are the first to lose their snow, and ground snow pack is shallower.